AF118553

Innovative Animal Manure Management for Environmental Protection, Improved Soil Fertility and Crop Production

Innovative Animal Manure Management for Environmental Protection, Improved Soil Fertility and Crop Production

Special Issue Editors

Kyoung S. Ro
Ariel A. Szogi
Gilbert C. Sigua

MDPI • Basel • Beijing • Wuhan • Barcelona • Belgrade

Special Issue Editors
Kyoung S. Ro
USDA Agricultural Research Service
USA

Ariel A. Szogi
USDA Agricultural Research Service
USA

Gilbert C. Sigua
USDA Agricultural Research Service
USA

Editorial Office
MDPI
St. Alban-Anlage 66
4052 Basel, Switzerland

This is a reprint of articles from the Special Issue published online in the open access journal *Environments* (ISSN 2076-3298) in 2019 (available at: https://www.mdpi.com/journal/environments/special_issues/animal_manure_management).

For citation purposes, cite each article independently as indicated on the article page online and as indicated below:

LastName, A.A.; LastName, B.B.; LastName, C.C. Article Title. *Journal Name* **Year**, *Article Number*, Page Range.

ISBN 978-3-03921-956-8 (Pbk)
ISBN 978-3-03921-957-5 (PDF)

Cover image courtesy of Kyoung S. Ro.

Contents

About the Special Issue Editors vii

Preface to "Innovative Animal Manure Management for Environmental Protection, Improved Soil Fertility and Crop Production" ix

Mindy J. Spiehs, Bryan L. Woodbury and David B. Parker
Ammonia, Hydrogen Sulfide, and Greenhouse Gas Emissions from Lab-Scaled Manure Bedpacks with and without Aluminum Sulfate Additions
Reprinted from: *Environments* **2019**, *6*, 108, doi:10.3390/environments6100108 1

John Loughrin and Nanh Lovanh
Aeration to Improve Biogas Production by Recalcitrant Feedstock
Reprinted from: *Environments* **2019**, *6*, 44, doi:10.3390/environments6040044 14

Kyoung S Ro, Mark A Dietenberger, Judy A Libra, Richard Proeschel, Hasan K. Atiyeh, Kamalakanta Sahoo and Wonkeun J Park
Production of Ethanol from Livestock, Agricultural, and Forest Residuals: An Economic Feasibility Study
Reprinted from: *Environments* **2019**, *6*, 97, doi:10.3390/environments6080097 24

Berta Riaño, Beatriz Molinuevo-Salces, Matías B. Vanotti and María Cruz García-González
Application of Gas-Permeable Membranes For-Semi-Continuous Ammonia Recovery from Swine Manure
Reprinted from: *Environments* **2019**, *6*, 32, doi:10.3390/environments6030032 38

María Soto-Herranz, Mercedes Sánchez-Báscones, Juan Manuel Antolín-Rodríguez, Diego Conde-Cid and Matias B. Vanotti
Effect of the Type of Gas-Permeable Membrane in Ammonia Recovery from Air
Reprinted from: *Environments* **2019**, *6*, 70, doi:10.3390/environments6060070 51

Ariel A. Szogi, Paul D. Shumaker, Kyoung S. Ro and Gilbert C. Sigua
Nitrogen Mineralization in a Sandy Soil Amended with Treated Low-Phosphorus Broiler Litter
Reprinted from: *Environments* **2019**, *6*, 96, doi:10.3390/environments6080096 63

Karamat R. Sistani, Jason R. Simmons, Marcia Jn-Baptiste and Jeff M. Novak
Poultry Litter, Biochar, and Fertilizer Effect on Corn Yield, Nutrient Uptake, N_2O and CO_2 Emissions
Reprinted from: *Environments* **2019**, *6*, 55, doi:10.3390/environments6050055 76

Philip J. Bauer, Ariel A. Szogi and Paul D. Shumaker
Fertilizer Efficacy of Poultry Litter Ash Blended with Lime or Gypsum as Fillers
Reprinted from: *Environments* **2019**, *6*, 50, doi:10.3390/environments6050050 90

Gilbert C. Sigua, Jeff M. Novak, Don W. Watts, Jim A. Ippolito, Thomas F. Ducey, Mark G. Johnson and Kurt A. Spokas
Phytostabilization of Zn and Cd in Mine Soil Using Corn in Combination with Biochars and Manure-Based Compost
Reprinted from: *Environments* **2019**, *6*, 69, doi:10.3390/environments6060069 102

Jeffrey M. Novak, Gilbert C. Sigua, Thomas F. Ducey, Donald W. Watts and Kenneth C. Stone
Designer Biochars Impact on Corn Grain Yields, Biomass Production, and Fertility Properties of a Highly-Weathered Ultisol
Reprinted from: *Environments* **2019**, *6*, 64, doi:10.3390/environments6060064 . . . . **121**

Thomas F. Ducey, Diana M. C. Rashash and Ariel A. Szogi
Differences in Microbial Communities and Pathogen Survival Between a Covered and Uncovered Anaerobic Lagoon
Reprinted from: *Environments* **2019**, *6*, 91, doi:10.3390/environments6080091 . . . . **136**

Thomas F. Ducey, Diana M. C. Rashash and Ariel A. Szogi
Correction: Ducey et al. Differences in Microbial Communities and Pathogen Survival Between a Covered and Uncovered Anaerobic Lagoon. *Environments*, 2019, *6*, 91
Reprinted from: *Environments* **2019**, *6*, 109, doi:10.3390/environments6100109 . . . . **149**

About the Special Issue Editors

Kyoung S. Ro is a research environmental engineer for the USDA-ARS Coastal Plains, Soil, Water, and Plant Research Center in Florence, South Carolina. He graduated with Ph.D. and M.S. in Civil Engineering from University of California at Los Angeles and B.S. in Chemical Engineering from University of California at Berkeley. His current research focus areas are state-of-art monitoring and control measures for fugitive gas emission from agricultural activities and innovative agro-environmental-soil applications of biochar/hydrochar. He has published more than 170 technical articles with 4000+ citations and organized several international symposia on fugitive gas emission, biofuels, and hydrothermal carbonization.

Ariel A. Szogi is the Research Leader of the USDA-ARS Coastal Plains, Soil, Water, and Plant Research Center in Florence, South Carolina. He graduated with a Ph.D., in Soil Sciences from Louisiana State University and M.Sci. in Soil and Water Management from Wageningen University. His research is focused on development of practical solutions to soil and water environmental problems related to land disposal of wastes from intensive animal and agricultural production. He has been instrumental in providing leadership and key expertise on new manure treatment technologies worldwide. His research has been documented in over 250 technical publications including 18 patents. He has conducted cooperative work across the United States, Asia, Europe, and Latin America. He is an elected Fellow of the American Society of Agronomy.

Gilbert C. Sigua is a Research Soil Scientist at the USDA-ARS Coastal Plains Soil, Water, and Plant Research Center in Florence, SC. He graduated with a Ph.D., in Soil Environmental Chemistry from Louisiana State University and M.S. in Soil Chemistry and Biochemistry from the University of Arkansas. His research program focuses on both the short-term and long-term solutions to enhancing agricultural and environmental sustainability and improving water and nutrient management in humid region. Gilbert is a nationally and internationally recognized expert in his field because of his work on agronomic, soil quality, water quality and environmental management research as evidenced by his various international projects in Asia, South America, and Africa. His research has been documented in over 180 technical publications. He is an elected Fellow of the American Society of Agronomy, Japan Society for Promotion of Science Fellow, and Soil Science Society of America.

Preface to "Innovative Animal Manure Management for Environmental Protection, Improved Soil Fertility and Crop Production"

Various innovative technologies to produce biogas, ethanol, and ammonia from livestock residuals are introduced in this issue. Loughrin and Lovanh [1] reported that supplying small amounts of air into an anaerobic digester increased biomass production. Supplying up to 800 mL/d air to a 133 L poultry litter slurry increased biogas production by 73%; however, aeration at 2000 mL/d decreased biogas production by 19%. The research findings suggest that appropriate microaeration rates must be carefully determined in order to achieve optimal biogas production rates. Ro et al. [2] preliminarily evaluated the economic feasibility of producing ethanol from agricultural, livestock, and forest residuals using commercially available gasification-synthesis gas fermentation technologies. A preliminary cost analysis of the integrated system was made for two cases: the regional scale of a 50 million-gallon (50 MGY or 189,271 m3)-per-year facility and a co-op-scale (1–2 MGY) facility. The minimum ethanol selling prices (MESP) depend heavily on the facility size and feedstock costs. The MESP for the 50 MGY facility were significantly lower and comparable to current gasoline prices ($2.24–$2.96 per gallon or $0.59–$0.78 per liter) for the low-value feedstocks such as wood, wheat, straw blended with dewatered swine manure, and corn stover.

Not only can biogas and ethanol be produced from animal manures, but ammonia can also be directly extracted from swine manure using membrane technology. Riano et al. [3] reported up to 90% recovery of total ammonia nitrogen (TAN) from swine manure using a gas-permeable membrane. They suggested that semicontinuous gas-permeable membrane technology may have a great potential for TAN recovery from animal manure. In addition, the ammonia recovery effectiveness of three types of gas-permeable membranes were investigated by Soto-Herranz et al. [4] These membranes were all made of expanded polytetrafluoroethylene (ePTFE) but with different diameters (3.0–8.6 mm), polymer densities (0.49–1.09), air permeability (2–40 L min^{-1} cm^2), and porosity (5.6–21.8%). While the ammonia recovery yields were not affected by the large differences in density, porosity, air permeability, and wall thickness, use of membranes with larger diameter and corresponding larger surface area yielded higher ammonia recovery. Higher fluid velocity of the circulating acidic solution significantly increased ammonia recovery.

Considering the worldwide demand and increasing costs of synthetic fertilizers, the utilization of animal manures and their byproducts as sources of plant nutrients is regarded as a favorable alternative to improve farm income while restoring soil fertility and protecting the environment. Three articles in this Special Issue address the potential environmental impact of the use of manure byproducts as soil amendments regarding their efficacy as fertilizers, application distribution, and emissions of ammonia and greenhouse gases. In the first article, Bauer et al. [5] addressed the fertilizer value of incinerated poultry litter ash along with optimal land application methodologies. They evaluated the use of calcitic lime and flue gas desulfurization gypsum (FGDG) as potential fillers for land application of poultry litter ash. Application of ash alone or with fillers significantly increased soil extractable P and K levels above unamended controls by 100% and 70%, respectively. A field application distribution test suggested that uniform distribution of ash alone or with fillers is feasible with commercial spinner disc fertilizer applicator. The two other articles report the effect of acidification of manure as a strategy to lower nitrogen losses due to ammonia volatilization.

Spieh et al. [6] tested the acidifying effect of alum (aluminum sulfate) used as an amendment to abate emissions of ammonia losses along with greenhouse gases (methane and nitrous oxide), and hydrogen sulfide gas from cattle manure bedpacks. Their results indicated that an application of 10% alum is needed to effectively diminish ammonia emissions. While nitrous oxide emissions were not affected by the alum treatment, methane and hydrogen sulfur increased with the addition of alum. As a second alternative to conserve nitrogen, Szogi et al. [7] studied the use of low-phosphorus broiler litter, a byproduct of the Quick Wash process designed to manage the surplus of nitrogen and phosphorus prior to soil application of treated low-phosphorus litter, which appears as an option for slow mineral nitrogen release and abatement of ammonia and nitrous oxide soil losses.

Anaerobic lagoons are a conventional of manure treatment for confined swine production systems in the southeastern United States. Using a synthetic cover, these lagoons can be modified to capture the emission of ammonia and other malodorous compounds. Ducey et al. [8] assessed the potential of these covers to alter lagoon microbial communities under the assumption that alterations in the physicochemical makeup due to use of a lagoon cover can impact the biological properties, most notably, the pathogenic populations. Their results show the addition of a cover had a significant impact on fecal coliform and E. coli levels, resulting in increased counts with respect to the uncovered, likely to due to a reduction in solar irradiation. From their microbial community composition—in which 200 bacterial families were identified—they concluded that synthetic covers play a role in changing the lagoon microclimate, impacting the lagoon's physicochemical and biological properties.

The recycling and use of raw materials from human-generated animal wastes are some of the environmental challenges that we face today. The promotion of innovative and appropriate technologies is necessary to achieve sound and sustainable animal manure management. Biochar production using pyrolysis technology can utilize most animal manure and many other recycled organics. Biochar is the solid product that results from pyrolysis of organic materials. Recycling of animal manure as a low-cost organic fertilizer has resulted in a favorable effect on improving the yield of a variety of crops and promoted the ecological and environmental functions of soils. The organic matter contents of pyrolyzed and/or composted animal manures is considerably high, and its addition to agricultural soils improves the soil physical, chemical, and biological properties.

Studies included in this volume have highlighted the effectiveness of pyrolyzed and composted animal manures as soil amendments that improve soil conditions and increase the agronomic values of these soils. The incorporation of these organic amendments can also improve the quality of contaminated mine soils and makes it possible for vegetation to be established. Karamat et al. [9] conducted a field plot study to investigate the impact of biochar and poultry litter, alone or in combination, on corn biomass, grain yield, nutrient uptake, and greenhouse gas emission for three growing seasons in Bowling Green, Kentucky. They reported that poultry litter application alone produced a significantly greater corn yield than biochar application, but similar to chemical fertilizer application. Addition of fertilizer or poultry litter had a positive effect on reducing N2O and CO2 fluxes compared to fertilizer or poultry litter application alone. Additionally, there was a slight increase in grain yield for each year following biochar application and when biochar was mixed with poultry litter or fertilizers. Novak et al. [10] reported that designer biochars were able to improve important fertility properties in the sandy Goldsboro soil located in Florence, South Carolina. Despite the noted improvement in soil fertility, corn grain and biomass yields were not significantly increased. The lack of significant improvement in observed corn yields corroborates the results from other biochar field research project conducted in temperate regions. They concluded that despite the Goldsboro soil being extensively weathered, it is still possessed sufficient soil fertility traits that can produce satisfactory corn yields given good agronomic practices and timely rainfall.

The work of Sigua et al. [11] reported in this volume has underscored the favorable advantage of mixing biochar with manure-based compost on enhancing the shoot and root biomass and nutritional uptake of corn grown in mine soils with heavy metal contaminations. The greatest total corn biomass was from soils treated with manure-based biochars (i.e., poultry litter, beef cattle manure) and the least total biomass was from wood-based biochar (lodge pole pine). Results of their study showed that the incorporation of biochar enhanced phytostabilization of Cd and Zn, with concentrations of water-soluble Cd and Zn lowest in soils amended with manure-based biochars while improving the biomass productivity of corn. They concluded that the phytostabilization technique, when combined with the biochar and manure-based compost application, has potential for the remediation of heavy metal-polluted soils.

Kyoung S. Ro, Ariel A. Szogi, Gilbert C. Sigua
Special Issue Editors

environments

MDPI

Article

Ammonia, Hydrogen Sulfide, and Greenhouse Gas Emissions from Lab-Scaled Manure Bedpacks with and without Aluminum Sulfate Additions

Mindy J. Spiehs [1,*], Bryan L. Woodbury [1] and David B. Parker [2]

1 USDA Agricultural Research Service, Meat Animal Research Center, Spur 18D, Clay Center, NE 68933, USA; bryan.woodbury@usda.gov

2 USDA Agricultural Research Service, Conservations and Production Research Laboratory, Bushland, TX 79012, USA; david.parker@usda.gov

* Correspondence: mindy.spiehs@usda.gov

Received: 26 July 2019; Accepted: 18 September 2019; Published: 20 September 2019

Abstract: The poultry industry has successfully used aluminum sulfate (alum) as a litter amendment to reduce NH_3 emissions from poultry barns, but alum has not been evaluated for similar uses in cattle facilities. A study was conducted to measure ammonia (NH_3), greenhouse gases (GHG), and hydrogen sulfide (H_2S) emissions from lab-scaled bedded manure packs over a 42-day period. Two frequencies of application (once or weekly) and four concentrations of alum (0, 2.5, 5, and 10% by mass) were evaluated. Frequency of alum application was either the entire treatment of alum applied on Day 0 (once) or 16.6% of the total alum mass applied each week for six weeks. Ammonia emissions were reduced when 10% alum was used, but H_2S emissions increased as the concentration of alum increased in the bedded packs. Nitrous oxide emissions were not affected by alum treatment. Methane emissions increased as the concentration of alum increased in the bedded packs. Carbon dioxide emissions were highest when 5% alum was applied and lowest when 0% alum was used. Results of this study indicate that 10% alum is needed to effectively reduce NH_3 emissions, but H_2S and methane emissions may increase when this concentration of alum is used.

Keywords: ammonia; beef; bedding; carbon dioxide; greenhouse gas; hydrogen sulfide; methane; nitrous oxide

---

1. Introduction

Ammonia (NH_3) is one of many common by-products of livestock production. The nitrogen (N) content in animal feeds is often fed in excess of the animals' nutrient requirements, resulting in surplus N that is excreted in the urine and feces of the animal. Stowell reported that only an estimated 10–30% of N that is consumed by cattle is utilized by the animal for growth, reproduction, milk production, and maintenance needs, with the remainder being excreted [1]. Nitrogen is an essential nutrient for plants, and the N retained in animal waste or on the feedlot surface can be used to fertilize cropland. Unfortunately, N losses as NH_3 from beef cattle feedlots can be quite high. Research conducted at Texas feedlots found 68% N loss as NH_3 in summer and 36% during winter months [2]. Similar research conducted in Nebraska feedlots found 51–63% N loss as NH_3 during summer months and 35–41% during winter months [3–5]. Nitrogen is primarily excreted through urine, with upward of 97% of urinary N in the form of urea [6,7]. When urine and feces are exposed to each other, the urea in urine is rapidly converted to NH_3 via the enzyme urease, which is found in feces [8]. This reaction is modulated by pH and temperature, with greater volatilization during the summer months compared to winter months [3–5]. The ideal pH for NH_3 volatilization is 7 to 10 [9], whereas pH ≤ 6.5 will result in little NH_3 volatilization [10]. This is a function of the pKa, with ammonium (NH_4) being less volatile than NH_3. Consequently, amendments that lower the pH of the beef feedlot surface material may reduce N volatilization and retain N in the manure to be used for fertilizer. This would be especially beneficial during the warm summer months.

Aluminum sulfate (alum) has been used successfully in the poultry industry to lower NH_3 emissions for the past decade [11,12]. Alum lowers the pH of the litter to bind N as NH_4 and prevent volatilization. The typical application rate is 5–10% of the weight of the litter [13–15]. Ammonia fluxes from alum-treated litter have been reported to be 70% lower than untreated litter [11]. Litter treated with alum also has a higher N content than untreated litter, which increases the fertilizer value of the litter [12]. Phosphorus (P) leaching is also reduced when the litter is land applied, due to a lower soluble P content in the alum-treated litter compared to untreated litter. We hypothesized that NH_3 volatilization would be lower from a manure and bedding mixture treated with alum compared to an untreated mixture of manure and bedding. However, we were unsure if alum would affect greenhouse gas and hydrogen sulfide (H_2S) volatilization from the bedpacks treated with alum.

2. Materials and Methods

A two × four factorial study with repeated measures over time was conducted to examine the effects of alum addition on gas emissions from manure bedded packs. Two dose frequencies (once or weekly) and four concentrations of alum (0, 2.5, 5, or 10% alum by mass) were examined. Thirty-two laboratory-scaled simulated bedded packs were constructed and maintained for 42 days, as previously described [16–21]. The study was conducted twice (Period 1 and Period 2) for a total of 64 experimental units. Briefly, bedded packs were constructed using plastic containers that were 0.5 m high and had a diameter of 0.38 m. Each container had six 1 cm holes equally spaced around the circumference of the container, approximately 5 cm from the top of the container, to serve as air inlets [21]. Corn stover was used as the bedding material. Three times per week, urine and feces were added to the bedded packs. The bedded packs were stirred slightly at each addition to represent hoof action on the bedded packs. The bedded packs were housed in four environmental chambers [22] that were maintained at an ambient temperature of 18 °C with a dew point of 12 °C throughout the study. This temperature and dew point were selected to represent conditions during a moderate season in a barn [23], like spring or fall, and have been used in previous studies [16–21]. There were eight bedded packs per chamber.

Once weekly, air samples were collected for 18 minutes per bedded pack using dynamic flux chambers. The flux chambers were stainless steel hemispherical flux chambers that were 7 L with a surface area of 640 cm^2 [24,25]. Inside the headspace of the chamber was a 40 mm, 12 V axial-flow fan moving approximately 130 L min^{-1}. The fan was suspended in the center of the headspace approximately 70 mm above the bedded pack surface. The fan airflow direction was from the surface to the top of the chamber. Rubber skirts that were 61 cm square and made of soft, elastic rubber with 22.9 cm diameter holes cut in the center were fit over each flux chamber to form a seal on the top of the plastic container when the flux chamber was placed on the plastic container for air sampling [21]. A 0.64 cm inert tubing was attached to the flux chambers using inert compression fittings [21]. The inert tubing was attached to the gas sampling manifold that fed into the air sampling equipment [21]. The gas sampling system was controlled by a 24 V Programmable Logic Relay, which signaled multi-positional three-way solenoids to open and close one of eight air inlet lines on the gas sampling manifold [21]. One line was opened at a time to allow for individual air sampling from each bedded pack [21]. Ambient air from the room was flushed through the tubing at a rate of 5 L min^{-1} for 30 min [21]. After the 30 min flush period, sampling was conducted for 18 minutes per bedded pack. Air samples were analyzed for H_2S using a Thermo Fisher 450i Hydrogen Sulfide/Sulfur Dioxide/Combined Sulfur Pulsed Fluorence gas analyzer (Thermo Fisher Scientist, Waltham, MA, USA). A Thermo Fisher 17i Ammonia Chemiluminescent gas analyzer was used to measure NH_3 (Thermo Fisher Scientist, Waltham, MA, USA). Methane was measured using a Thermo Fisher 55i Direct Methane and Non-methane Hydrocarbon Backflush Gas Chromatograph gas analyzer (Thermo

Fisher Scientific, Waltham, MA, USA). Nitrous oxide (N_2O) and carbon dioxide (CO_2) were measured using an Innova 1412 Photoacoustic gas monitor (LumaSense Technologies, Santa Clara, CA, USA).

The temperature and pH of each bedded pack were measured at each sampling date. A grab sample of approximately 10–15 g was collected from each bedded pack and placed in a 50 mL plastic conical. Samples were diluted 1:2 on a mass basis with distilled water. A pH/mV/temperature meter (IQ150, Spectrum Technologies, Inc., Plainsfield, IL, USA), calibrated with buffers pH 4 and 7, was then used to determine pH. Pack temperature was measured approximately 7.6 cm below the surface of the simulated bedded pack using the same meter. At the beginning and end of the study, a sample of each simulated bedded pack was collected and analyzed for dry matter (DM), total N, total phosphorus (P), total potassium (K), and total sulfur (S). Dry matter was determined by weighing samples before and after drying at 100 °C in a forced-air oven for 24 h. Samples collected from the simulated bedded packs were dried, ground through a 1 mm screen, and sent to a commercial laboratory (Ward Laboratory, Inc., Kearney, NE) for N [26], P, S [27] analysis.

Treatments were applied on Day 0 and measurements were recorded once weekly for the following six weeks (Day 42). The treatments included 0, 2.5, 5, and 10% alum applied as either a whole treatment on Day 0 or 16.6% of the total alum volume applied each week for six weeks. The quantity of alum added was based on the expected total mass of the bedded packs at the end of the 42-day study. Treatments were selected based on previous research in the poultry industry, which indicated 5–10% alum by mass was adequate to reduce NH_3 volatilization [13–15]. As reported by Moore and co-workers [13], a series of studies has consistently demonstrated that pH begins to rise and NH_3 emissions begin to increase about four weeks post application when used in poultry facilities. This study was designed to slightly exceed that four-week threshold, to determine the maximum length of time NH_3 emissions could be suppressed from cattle bedded packs.

For all measurements, Day 0 samples were collected prior to any urine, feces, or alum being added to the bedding material. The reason for doing this was to determine baseline emissions from the corn stover bedding independently of urine, feces, or alum. Previous studies with alum had all been conducted in poultry barns that use wood-based bedding material. We wanted to make sure that corn stover alone was not contributing to high gaseous emissions. In most cases, these measurements were essentially zero.

Data were analyzed as a 4×2 factorial with repeated measures in time using the MIXED procedure of SAS (SAS Institute, Cary, NC, USA). The model included the effects of period, alum concentration, dose frequency, day, and all interactions. Covariate structure was modeled to get lowest Akaike information criteria (AIC) value. Several structures were tested and the compound symmetry covariate structure provided the lowest AIC value. When significant differences were detected, the least square means were calculated using Fisher's least significant differences. The nutrient data was analyzed as a 4×2 factorial using the MIXED procedure of SAS (SAS Institute, Cary, NC, USA) to test the effects of alum concentration, dose frequency, and the interaction of the two variables. Bedded pack was the experimental unit in all analyses, and differences were considered significant when $p < 0.05$.

3. Results and Discussion

3.1. Temperature and pH

The temperature of the bedded packs ranged from 19.1 to 21.7 °C and did not differ amongst bedding packs at any point during the 42-day study (Table 1). These temperatures were within the range of previous studies using a similar lab-scaled design [16,20], but less than pack temperatures measured by Ayadi [28] when using environmental chambers set at 40 °C. The average temperature of the bedded pack in four commercial facilities was 25.7 °C but ranged from 15 to 29 °C depending on the season [23]. Although no lab-scale system can perfectly simulate the environment in a commercial facility, the lab-scale simulated bedded packs have consistently produced physical characteristics within the ranges measured in commercial facilities and allow researchers a small-scaled tool that can be used for initial evaluation of multiple treatments.

Table 1. Mean pH and temperature (°C) of bedded pack material on Day 0–42 following the addition of alum.

pH								
Dose Frequency	Alum	Day						
		0	7	14	21	28	35	42
Once	0%	7.06	8.49 ^{a}	8.18 ^{a}	8.03 ^{a}	7.97 ^{a}	7.81 ^{a}	7.79 ^{b}
Weekly	0%	7.04	8.51 ^{a}	8.27 ^{a}	8.10 ^{a}	7.76 ^{a,b}	7.91 ^{a}	7.82 ^{b}
Once	2.5%	7.05	8.43 ^{a,b}	8.16 ^{a,b}	8.14 ^{a}	7.84 ^{a,b}	7.98 ^{a}	7.96 ^{a}
Weekly	2.5%	6.95	8.44 ^{a}	8.19 ^{a}	8.04 ^{a}	7.77 ^{a,b}	7.85 ^{a}	7.66 ^{b,c}
Once	5%	7.00	8.03 ^{c}	8.14 ^{a}	8.00 ^{a}	7.75 ^{a,b}	7.85 ^{a}	7.77 ^{b}
Weekly	5%	7.20	8.29 ^{a,b}	8.08 ^{a,b}	8.03 ^{a}	7.75 ^{a,b}	7.83 ^{a}	7.83 ^{b}
Once	10%	7.00	6.92 ^{d}	7.93 ^{b}	7.97 ^{a,b}	7.68 ^{b}	7.84 ^{a}	7.70 ^{b,c}
Weekly	10%	7.17	8.21 ^{a,b}	8.17 ^{a,b}	7.62 ^{b}	7.64 ^{b}	7.65 ^{b}	7.60 ^{c}
Temperature								
Dose Frequency	Alum	Day						
		0	7	14	21	28	35	42
Once	0%	20.0	19.1	20.2	21.2	21.7	21.1	21.6
Weekly	0%	20.0	19.1	20.0	21.4	21.3	21.1	21.6
Once	2.5%	20.0	19.1	20.0	21.3	21.4	20.9	21.6
Weekly	2.5%	20.0	19.2	20.1	21.1	21.4	20.9	21.6
Once	5%	20.0	19.1	19.9	21.0	21.1	20.9	21.5
Weekly	5%	20.0	19.2	19.7	21.1	21.1	20.7	21.5
Once	10%	20.0	19.3	20.0	21.0	21.1	20.8	21.7
Weekly	10%	20.0	19.4	20.4	21.2	21.4	21.1	21.7

Within a column, different superscripts indicate significant differences between treatments ($p < 0.05$).

The pH of the bedded packs was also within the range of previous lab-scaled studies (6.2 to 9.0; [16,20,28]) and data collected from commercial facilities (7.5–8.0; [23]). Previous studies using 10% alum in poultry litter reported a litter pH ranging from approximately 7.5 to 8.0 in untreated litter to approximately 5.75 immediately after a 10% alum treatment [13]. The pH began to gradually increase and reached approximately 7.5 four weeks after application [13]. In the current study, the pH of the untreated bedded packs was similar to untreated litter, but 10% alum was able to lower the pH of poultry litter to 5.75, while the 10% treatment applied in one dose only lowered the pH of the bedded pack to 6.92. Further studies will be warranted to determine if a higher dosage of alum may be necessary to sufficiently lower pH. As expected, the pH of the bedded packs changed when alum was added. Initial measurements from the bedded pack taken on Day 0 were collected immediately before adding the alum and did not show any significant differences between the treatments. On Day 7, the pH of the bedded packs that received the 10% alum treatment in one dose was lower ($p < 0.01$) than the pH of bedded packs that received all other treatments. The bedded packs that received the 5% alum treatment in one dose had the second lowest pH, followed by the 10% weekly treatment, 5% weekly treatment, and 2.5% dose administered at once. Expectedly, the largest concentration of alum administered in one dose was the most effective in lowering the initial pH. This low pH was sustained through Day 14, but by Day 21, only the bedded packs receiving the 10% weekly alum treatment maintained a lower pH. Alum applied to poultry litter prior to a six-week grow-out was particularly effective at lowering pH during the first three to four weeks [11]. Therefore, it was not surprising that the alum would begin to lose efficacy around Day 21.

Beginning on Day 28, an overall decrease in pH of all bedded packs was observed, with the bedded packs containing the 10% alum treatment having significantly lower pH than those that did not receive alum. This decrease in bedded pack pH has been observed in previous studies using this experimental design, although the decrease did not happen until Day 35 in previous studies [17,19]. It is likely caused by an increase in volatile fatty acid (VFA) production as the manure mixture ages, as demonstrated by other researchers [16,29,30]. On Day 35, only the bedded packs receiving the 10%

alum weekly dose had a significantly lower pH. At the end of the study, the bedded packs with the 10% weekly alum treatment had the lowest pH, which did not differ statistically from the pH of the bedded packs receiving the 10% once treatment and the 2.5% weekly treatment, with all other bedded packs having a higher pH. It is unclear why the bedded pack with the 2.5% weekly treatment had such a significant drop in pH between Day 35 and Day 42.

3.2. Ammonia

Ammonia flux was significantly lower for bedded packs that received the 10% treatment in one dose compared to any other treatment at Day 7 (Figure 1). The bedded packs that received 2.5% alum had significantly higher NH_3 flux than those with 10% alum administered in one dose, but significantly lower than all other treatments on Day 7. At Day 14, the NH_3 fluxes from bedded packs that received both 10% alum treatments were not different from one another, but were significantly lower than other treatments, with those receiving the 2.5% and 5% alum in one dose having significantly higher NH_3 fluxes than the 10% treatments, but significantly lower NH_3 fluxes than the other treatments. From Day 21 until the end of the study, only the bedded packs with the 10% alum administered in weekly doses consistently maintained a significantly lower NH_3 flux than the bedded packs that were not treated with alum.

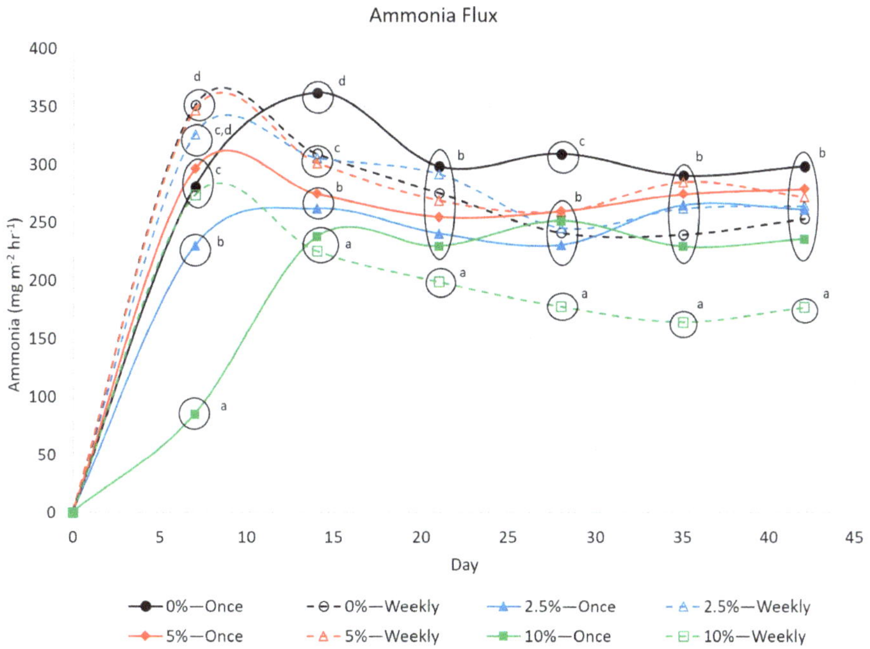

Figure 1. Average ammonia flux (mg m^{-2} h^{-1}) for weekly measurements taken over 42 days for bedded packs treated with 0, 2.5, 5, and 10% alum, either in one dose or in weekly doses. Alum × Dose × Day interaction $p < 0.01$, Pooled standard error of the mean = 8.19. Clusters with different letters indicate significant differences $p < 0.05$.

Ammonia flux followed the pack pH. On Day 7, the bedded packs that received the single dose of 10% alum had statistically lower NH_3 emission than all other bedded packs (Figure 1). This corresponded to the low Day 7 pH for the bedded pack that received a single dose of 10% alum. At Day 14, the pH of both 10% alum treatments, the 2.5% alum treatment that was administered once, and the 5% alum treatment that was administered once had the lowest pH values. The NH_3 emissions from these four treatments were also lowest on Day 14. By Day 21 and throughout the remainder of the study, the pHs of the bedded packs that were administered a weekly dose of 10% alum were lowest, which corresponded to the low NH_3 emissions for bedded packs that received weekly doses with the 10% alum concentration. Bedded packs that received weekly allotments of the 10% alum treatment had 73% of the NH_3 emissions of the bedded packs that did not receive an alum treatment.

3.3. Greenhouse Gases

There was a significant alum concentration × dose frequency × day effect for CO_2 flux (Figure 2). However, in general, the CO_2 flux from bedded packs receiving the 5% alum dose, regardless of whether it was administered weekly or in one dose, had the highest CO_2 flux, with the bedded packs that received no alum having the lowest CO_2 flux. It is unclear why the bedded packs receiving the 5% alum treatments produced the highest CO_2 flux. A reduction in pH could affect the concentration of gases near the surface of the bedded pack and would favor the emission of weak acid forming gases, such as CO_2 [31]. However, if the increase in CO_2 was purely pH-dependent, it would be expected that the bedded packs receiving the 10% alum treatment would have the highest CO_2 emission, not the bedded packs receiving the 5% alum treatments. The carbon dioxide flux of the bedded packs receiving the 10% alum in weekly allotments was 104% that of the untreated bedded packs, but did not differ statistically. Over the course of the 42-day experiment, the CO_2 flux increased for all bedded packs, regardless of treatments. This gradual increase in CO_2 flux is consistent with previous studies [17,19,28]. Carbon dioxide is produced during aerobic respiration and indicates biological activity in the bedded packs. As the packs matured, the decomposition of feces and bedding material began, which likely contributed to the increase in CO_2 emissions as the packs aged.

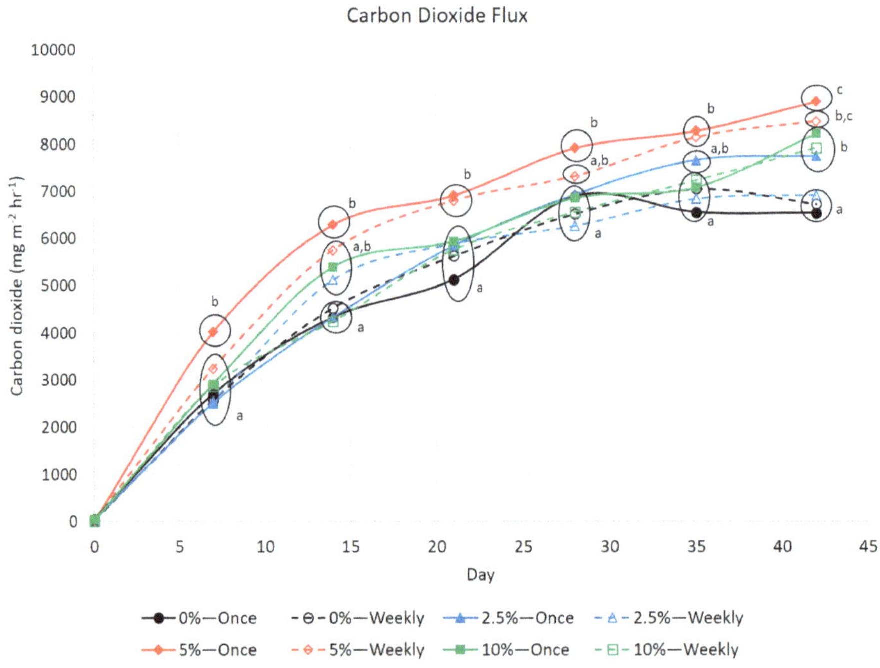

Figure 2. Average carbon dioxide flux (mg m^{-2} h^{-1}) for weekly measurements taken over 42 days for bedded packs treated with 0, 2.5, 5, and 10% alum, either in one dose or in weekly doses. Alum × Dose × Day interaction $p < 0.01$, Pooled standard error of the mean = 154.3. Clusters with different letters indicate significant differences $p < 0.05$.

Methane is produced during anaerobic decomposition of organic matter [32–35]. Methane-producing bacteria can survive in temperatures ranging from 15 to 60 °C, with an ideal temperature of 35–37 °C [36]. An ideal pH is between 6.7 and 7.5 [37,38]. Methane production was largely nonexistent until Day 21 of the study (Figure 3), which is consistent with previous studies conducted using simulated bedded packs [17,19]. While ideal conditions for CH_4 production were never achieved during this study, the weekly increases in bedded pack height as urine, feces, and bedding were added each week can be assumed to have made a more anaerobic environment towards the bottom of the bedded packs as the study progressed. Beginning at Day 28, the bedded packs that received the 10% alum treatments had the greatest CH_4 production, followed by those receiving both 5% alum treatments. This was likely due to the fact that the pH of the bedded packs containing 10% alum maintained a pH of 7.62–7.97 compared to higher pH values for the bedded packs that received all other treatments, particularly on Day 28, when CH_4 production peaked. The lower pH, while not ideal for methanogens, is certainly more conducive to the production of CH_4. For each concentration of alum applied to bedded packs, peak CH_4 production was significantly greater when the alum was applied in weekly allotments compared to one dose. Methane fluxes from the bedded packs that received the weekly dose of 10% alum were 274% of the flux from bedded packs that were not treated with alum. Methane production peaked at Day 28 and began to decline by Day 35.

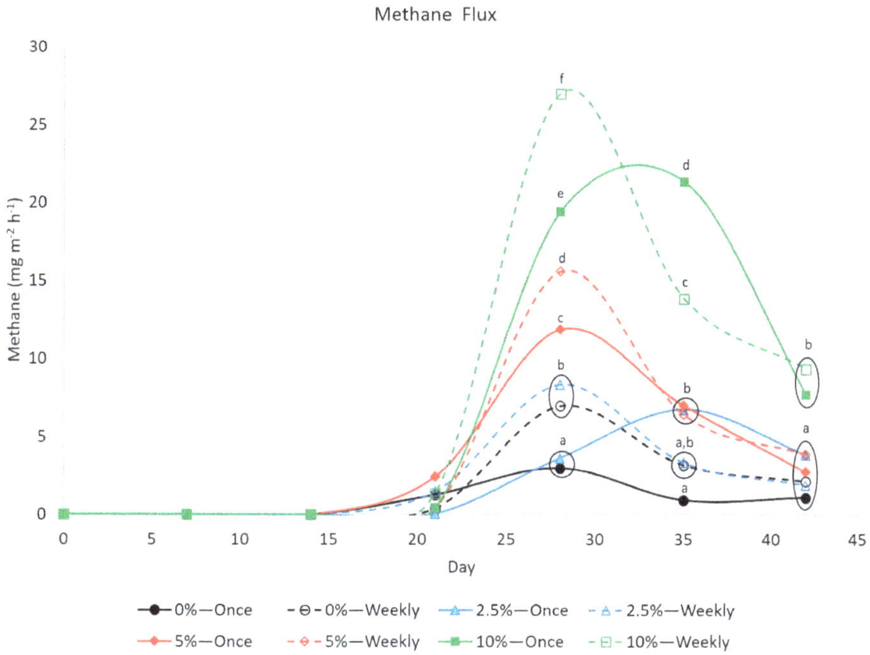

Figure 3. Average methane flux (mg m^{-2} h^{-1}) for weekly measurements taken over 42 days for bedded packs treated with 0, 2.5, 5, and 10% alum, either in one dose or in weekly doses. Alum × Dose × Day interaction $p < 0.01$, Pooled standard error of the mean = 0.76. Clusters with different letters indicate significant differences $p < 0.05$.

Nitrous oxide emissions from livestock facilities are of increasing concern for producers due to the global warming potential of N_2O being 296 times greater than that of CO_2 [39]. Nitrous oxide concentrations ranged from 0.01 to 0.69 mg m^{-2} h^{-1} and were highly variable across treatments throughout the study (Figure 4), which was similar to previous studies [17,19]. This variability over time may be attributed to the micro-environments on the surface of the bedded packs being disrupted with periodic stirring of the pack. When the pack is stirred, the anaerobic layer is disrupted and nitrifying and denitrifying processes may be altered as a result. Two peaks appeared, at Day 7 and Day 21, which is slightly different from previous studies, which had peaks at Day 14 and Day 42 [17–19]. At Day 7, bedded packs that received the 5% alum in one dose had significantly higher N_2O emissions and those that received the 10% weekly dose had significantly lower N_2O emissions, with all other treatments being similar. During the Day 21 peak, bedded packs that received the 2.5% weekly alum and 0% weekly dose had the highest N_2O emissions, with those receiving the 10% weekly dose once again having lower emissions. Overall, for the 42-day study, bedded packs treated with the 10% alum weekly treatment had lower N_2O emissions (Figure 5). Those given either 0% alum treatment, the 2.5% treatment in weekly increments, or the 5% alum in one dose had the highest N_2O loss, with the packs receiving the 10% alum treatment in one dose, the 5% alum treatment in weekly increments, and 2.5% alum treatment in one dose had intermediate N_2O emissions. Treating the bedded packs with weekly doses of 10% alum resulted in a flux 64% of that for the untreated bedded packs.

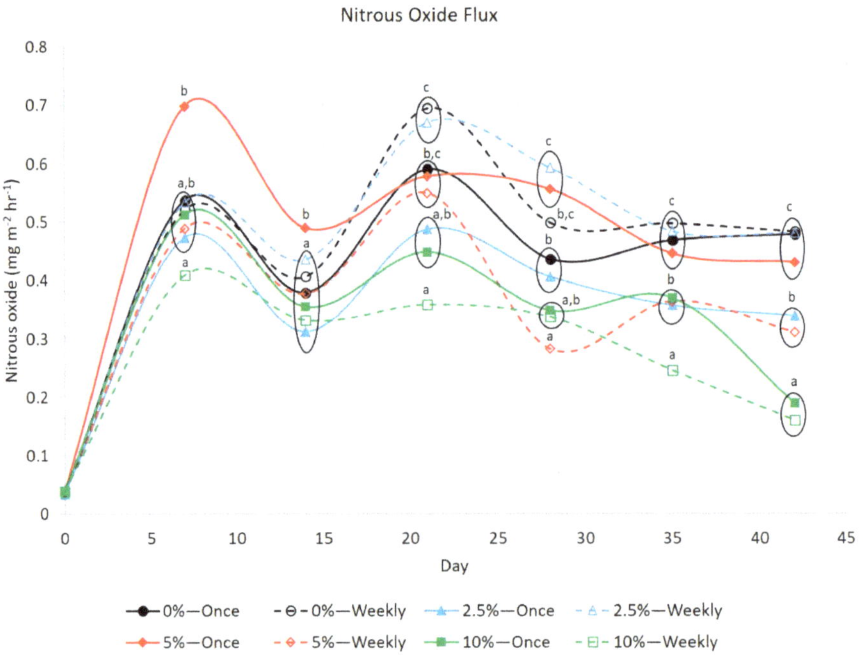

Figure 4. Average nitrous oxide flux (mg m^{-2} h^{-1}) for weekly measurements taken over 42 days for bedded packs treated with 0, 2.5, 5, and 10% alum either in one dose or in weekly doses. Alum × Dose × Day interaction $p < 0.01$, pooled standard error of the mean = 0.05. Clusters with different letters indicate significant differences $p < 0.05$.

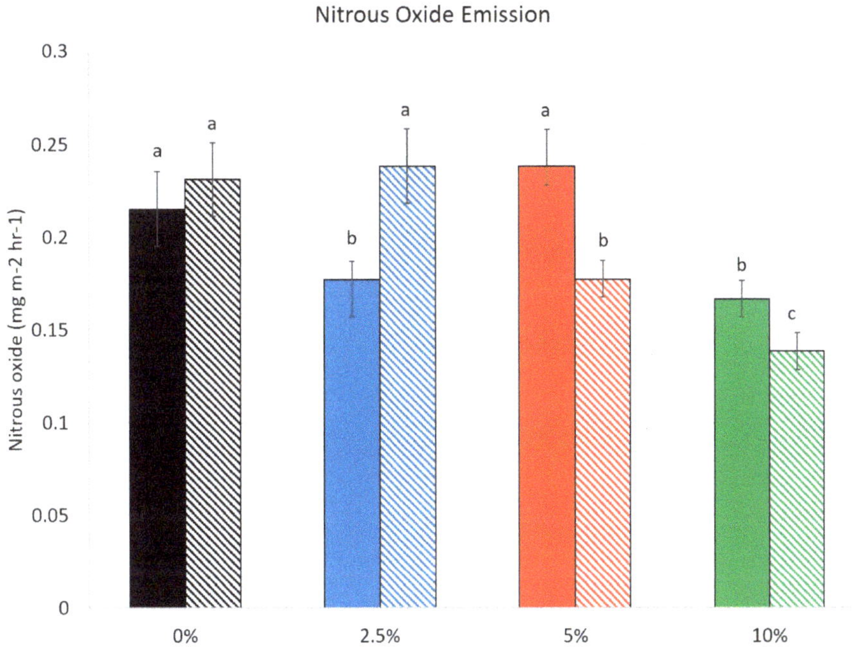

Figure 5. Nitrous oxide emission (mg m^{-2} h^{-1}) over 42 days for bedded packs treated with 0, 2.5, 5, and 10% alum either in one dose or in weekly doses. All treatments given in one dose are shown in solid colored bars and weekly increments are shown in striped bars for each treatment. Alum × Dose $p < 0.01$.

3.4. Hydrogen Sulfide

Hydrogen sulfide flux was initially very low, peaked at Day 14 and then declined throughout the study (Figure 6). In a previous study, hydrogen sulfide was not measured directly, but total reduced sulfides peaked at Day 14, declined, and then began to increase again at 42 days [16]. In a second study, H_2S continued to increase until the end of the study, at Day 42 [14]. Both of the previous studies included wood products which produced a secondary peak of H_2S, while the current study used only corn stover [17,19]. Sulfate-reducing bacteria are obligate anaerobes [40]. As the bedded packs aged, the material became more and more compacted, presumably creating an anaerobic environment under the surface of the bedded pack, which was conducive to the production of H_2S. At peak H_2S flux, the bedded packs that were treated with 10% alum in one dose had a significantly higher H_2S flux than any other bedded packs. The bedded packs that received the 10% alum in weekly doses had a H_2S flux not different to the bedded packs that received the 5% or 2.5% alum in one dose. From Day 14 to Day 28, bedded packs that received 10% alum treatments had the highest H_2S flux, but overall H_2S flux decreased as the packs aged. The hydrogen sulfide flux from the bedded packs treated with weekly doses of 10% alum was 168% of the flux from untreated bedded packs. The increase in H_2S emission from bedded packs that received the 10% weekly alum dose could be related to two factors: increase in available sulfate due to the addition of sulfur as alum or a change in pH. Bedded packs receiving 10% alum had the highest concentration of sulfur in the bedded pack material, and thus more available substrate for H_2S production. The release of gases from the bedded pack is also a function of the concentration of the gases near the surface in a non-ionized form, which can be easily affected by surface pH. A reduction in pH typically favors the emission of weak acid forming gases, such as CO_2

and H_2S [31]. While the H_2S flux did decrease over time, the higher flux observed when alum was used is a concern for producers using alum to control NH_3 emissions, as H_2S emissions may increase.

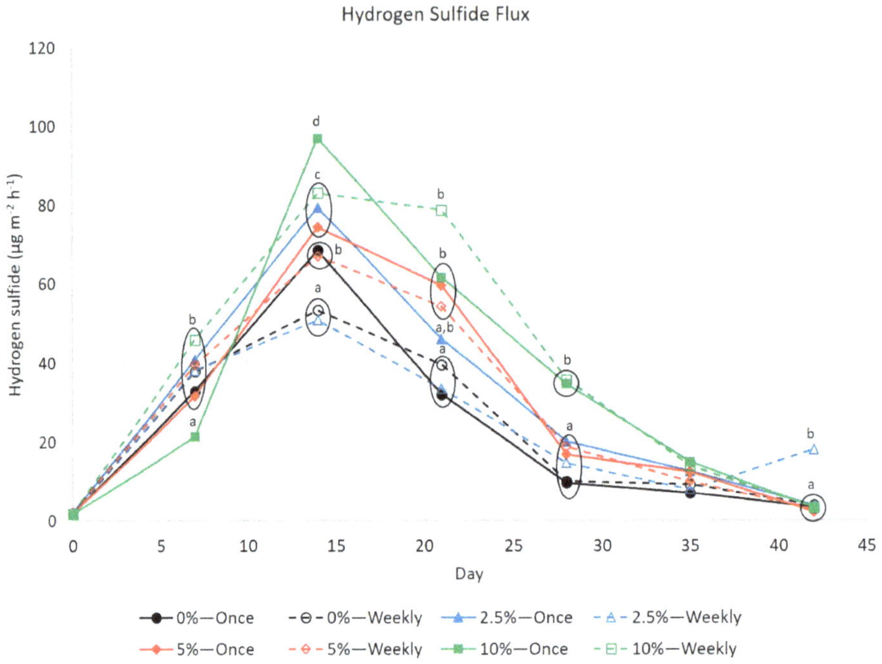

Figure 6. Average hydrogen sulfide flux (µg m^{-2} h^{-1}) for weekly measurements taken over 42 days for bedded packs treated with 0, 2.5, 5, and 10% alum either in one dose or in weekly doses. Alum × Dose × Day interaction $p < 0.01$, pooled standard error of the mean = 4.62. Clusters with different letters indicate significant differences $p < 0.05$.

3.5. Nutrient Composition

The average dry matter content of the bedded packs at the end of the study was 22.7% ± 1.3, which was similar to the dry matter content of corn stover bedded packs in previous studies conducted similarly [16,18,20] and slightly higher than the range of 16.9 ± 0.45 to 18.0 ± 0.33 measured over an 18-month period in commercial barns [23]. There was little difference in the nutrient composition of the bedded packs treated with the four concentrations of alum (Table 2). Because there were no interactions or differences due to the frequency of application, the table only includes nutrient concentrations for the main treatment effect of alum concentration. Phosphorus, potassium, and nitrogen content did not differ between all alum treatments. All bedded packs, regardless of treatment, had bedding, urine, and feces from a common source. Therefore, N input for all bedded packs was similar. With decreased NH_3 emission, it was expected that the bedded packs treated with 10% alum would have a higher total N concentration than the untreated bedded packs, due to decreased N volatilization as ammonia. However, this was not the case. Either N was emitted in another form, such as N_2 gas, that was not measured in this study, or N was retained in the pack but not detected using our analytical methods for total N detection. Perhaps some of the retained N was lost when the bedded pack material was dried prior to nutrient analysis. Another possibility for this anomaly is that the differences in cumulative N losses between the control and 10% alum treatments were small enough that they did not affect the total N concentration of the bedded pack materials. Sulfur increased with increasing levels of alum application. This was expected, as alum contains sulfur. This likely contributed to the increase in H_2S emissions from the bedded packs that received the 10% alum application.

Table 2. Average nutrient composition of the bedded packs at Day 42, following the addition of alum.

Alum	Phosphorus	Potassium	Sulfur	Nitrogen
	g kg^{-1} (DM)			
0%	0.90	3.40	5.09	26.76
2.5%	0.91	3.63	6.56	25.84
5%	0.93	3.55	8.01	26.28
10%	0.91	3.47	10.73	26.31
p-values				
Alum	0.9197	0.436	<0.01	0.5036
Frequency	0.4274	0.5431	0.5338	0.8204
Alum × Frequency	0.4269	0.5966	0.6204	0.6662

4. Conclusions

When alum was applied to a corn stover and manure mixture, it was necessary to use the 10% alum mixture to successfully lower NH_3 emissions compared to the corn stover manure bedpack that was not treated with alum. A one-time dose of 10% alum was only able to sustain the lowered NH_3 emissions for 14 days, but a total dose of 10% alum spread out in six equal weekly additions could maintain the reduced NH_3 emissions for up to 42 days. Nitrous oxide emissions were also reduced when 10% alum was administered weekly. However, CH_4 and H_2S emissions increased when the bedded packs were treated with 10% alum compared to bedded packs that did not receive an alum treatment. Producers need to be aware of the trade-offs of lowering one emission at the risk of increasing another and consider the benefits of lowering one emission relative to another. While effective, weekly additions of alum also present practical issues for livestock producers. It may be more effective to target areas of the pen most likely to emit NH_3. Producers may also consider less frequent applications, such as once every two or three weeks, but additional research is needed to determine the efficacy of this approach.

Author Contributions: Conceptualization, M.J.S., B.L.W., D.B.P.; Methodology, M.J.S.; Software, M.J.S.; Validation, M.J.S.; Formal Analysis, M.J.S.; Investigation, M.J.S., B.L.W.; Data Curation, M.J.S.; Writing–Original Draft Preparation, M.J.S.; Writing–Review & Editing, B.L.W., D.B.P.; Visualization, M.J.S..; Supervision, D.B.P.; Project Administration, M.J.S.; Funding Acquisition, M.J.S., B.L.W., D.B.P.

Funding: This project was funded by appropriated funds from the USDA. USDA is an equal opportunity provider and employer.

Acknowledgments: The author wishes to thank Alan Kruger, technician at USMARC, for help with data collection and lab analysis. The mention of trade names of commercial products in this article is solely for the purpose of providing specific information and does not imply recommendation or endorsement by the USDA.

Conflicts of Interest: The authors declare no conflict of interest.

References

1. Stowell, R. Ammonia loss and emission reporting: Considerations for cattle operations. *UNL BeefWatch*. 1 February 2018. Available online: http://newsroom.unl.edu/announce/beef/7498/42942 (accessed on 13 June 2018).
2. Todd, R.W.; Cole, N.A.; Clark, R.N.; Flesch, T.K. Ammonia emissions from a beef cattle feedyard on the Southern High Plains. *Atmos. Environ.* **2008**, *42*, 6797–6805. [CrossRef]
3. Bierman, S.; Erickson, G.E.; Klopfenstein, T.J.; Stock, R.A.; Shain, D.H. Evaluation of nitrogen and organic matter balance in the feedlot as affected by level and source of dietary fiber. *J. Anim. Sci.* **1999**, *77*, 1645–1653. [CrossRef] [PubMed]

4. Erickson, G.E.; Milton, C.T.; Klopfenstein, T.J. Dietary protein effects on nitrogen excretion and volatilization in open-dirt feedlots. In *Proceedings, Eighth International Symposium on Animal, Agricultural, and Food Processing Waste (ISAAF)*; American Society of Agricultural Engineers: St. Joseph, MI, USA, 2000.
5. Koelsch, R.; Erickson, G.; Homolka, M.; Luebbe, M. *Predicting Manure Nitrogen, Phosphorus, and Carbon Characteristics of Open Lot Systems*; ASABE Paper No. 1800946; ASABE: St. Joseph, MI, USA, 2018.
6. Bierman, S.; Klopfenstein, T.J.; Stock, R.; Shain, D. *Evaluation of Nitrogen, Phosphorus, and Organic Matter Balance in the Feedlot as Affected by Nutrition*; Beef Cattle Report MP66-A; University of Nebraska: Lincoln, NE, USA, 1996; pp. 74–76.
7. Van Horn, H.H.; Newton, G.L.; Kunkle, W.E. Ruminant nutrition from an environmental perspective: Factors affecting whole-farm nutrient balance. *J. Anim. Sci.* **1996**, *74*, 3082–3102. [CrossRef] [PubMed]
8. Hausinger, R.P. Metabolic versatility of prokaryotes for urea deposition. *J. Bacteriol.* **2004**, *186*, 2520–2522. [CrossRef] [PubMed]
9. Hartung, J.; Phillips, V.R. Control of gaseous emissions from livestock buildings and manure stores. *J. Agric. Eng. Res.* **1994**, *57*, 173–189. [CrossRef]
10. Rhoades, M.B.; Parker, D.B.; Cole, N.A.; Todd, R.W.; Caraway, E.A.; Auvermann, B.W.; Topliff, D.R.; Schuster, G.L. Continuous ammonia emission measurements from a commercial beef feedyard in Texas. *Trans. ASABE* **2010**, *53*, 1823–1831. [CrossRef]
11. Moore, P.A., Jr.; Daniels, T.C.; Edwards, D.R. Reducing phosphorus runoff and inhibiting ammonia loss from poultry manure with aluminum sulfate. *J. Environ. Qual.* **2000**, *29*, 37–49. [CrossRef]
12. Moore, P.A.; Edwards, D.R. Long-term effects of poultry litter, alum-treated litter, and ammonium nitrate on phosphorus availability in soils. *J. Environ. Qual.* **2007**, *36*, 163–174. [CrossRef]
13. Moore, P.A., Jr.; Daniels, T.C.; Edwards, D.R. Reducing phosphorus and improving poultry production with alum. *Poult. Sci.* **1999**, *78*, 692–698. [CrossRef]
14. Sims, J.T.; Luka-McCafferty, N.J. On-farm evaluation of aluminum sulfate (alum) as a poultry litter amendment: Effects on litter properties. *J. Environ. Qual.* **2002**, *31*, 2066–2073. [CrossRef]
15. Penn, C.; Zhang, H. Alum-Treated Poultry Litter as a Fertilizer Source. Oklahoma Cooperative Extension Service PSS-2254. 2017. Available online: http://pods.dasnr.okstate.edu/docushare/dsweb/Get/Document-5180/PSS-2254web13.pdf (accessed on 11 September 2019).
16. Spiehs, M.J.; Brown-Brandl, T.M.; Parker, D.B.; Miller, D.N.; Berry, E.D.; Wells, J.E. Effect of bedding materials on concentration of odorous compounds and *Escherichia coli* in beef cattle bedded manure packs. *J. Environ. Qual.* **2013**, *42*, 65–75. [CrossRef] [PubMed]
17. Spiehs, M.J.; Brown-Brandl, T.M.; Parker, D.B.; Miller, D.N.; Jaderborg, J.P.; DiCostanzo, A.; Berry, E.D.; Wells, J.E. Use of wood-based materials in beef bedded manure packs: 1. Effect on ammonia, total reduced sulfide, and greenhouse gas concentrations. *J. Environ. Qual.* **2014**, *43*, 1187–1194. [CrossRef] [PubMed]
18. Spiehs, M.J.; Brown-Brandl, T.M.; Berry, E.D.; Wells, J.E.; Parker, D.B.; Miller, D.N.; Jaderborg, J.P.; DiCostanzo, A. Use of wood-based materials in beef bedded manure packs: 2. Effect on odorous volatile organic compounds, odor activity value, *Escherichia coli*, and nutrient concentrations. *J. Environ. Qual.* **2014**, *43*, 1195–1206. [CrossRef] [PubMed]
19. Spiehs, M.J.; Brown-Brandl, T.M.; Parker, D.B.; Miller, D.N.; Berry, E.D.; Wells, J.E. Ammonia, total reduced sulfides, and greenhouse gases of pine chip and corn stover bedding packs. *J. Environ. Qual.* **2016**, *45*, 630–637. [CrossRef] [PubMed]
20. Spiehs, M.J.; Berry, E.D.; Wells, J.; Parker, D.B.; Brown-Brandl, T.M. Odorous volatile organic compounds, *Escherichia coli*, and nutrient composition when kiln-dried pine chips and corn stover bedding are used in beef bedded manure packs. *J. Environ. Qual.* **2017**, *46*, 722–732. [CrossRef] [PubMed]
21. Spiehs, M.J. Lab-scale model to evaluate odor and gas concentrations emitted by deep bedded pack manure. *J. Vis. Exp.* **2018**, *137*, e57332. [CrossRef]
22. Brown-Brandl, T.M.; Nienaber, J.A.; Eigenberg, R.A. Temperature and humidity control in indirect calorimeter chambers. *Trans. ASABE* **2011**, *54*, 685–692. [CrossRef]
23. Spiehs, M.J.; Woodbury, B.L.; Doran, B.E.; Eigenberg, R.A.; Kohl, K.D.; Varel, V.H.; Berry, E.D.; Wells, J. Environmental conditions in beef deep-bedded mono-slope facilities: A descriptive study. *Trans. ASABE* **2011**, *54*, 663–673. [CrossRef]
24. Miller, D.N.; Woodbury, B.L. A solid-phase microextraction chamber method for analysis of manure volatiles. *J. Environ. Qual.* **2006**, *35*, 2383–2394. [CrossRef]

25. Woodbury, B.L.; Miller, D.N.; Eigenberg, R.A.; Nienaber, J.A. An inexpensive laboratory and field chamber for manure volatile gas flux analysis. *Trans. ASABE* **2006**, *49*, 767–772. [CrossRef]
26. Watson, M.; Wolf, A.; Wolf, N. Total nitrogen. In *Recommended Methods of Manure Analysis*; Publication No. A3769; University of Wisconsin Cooperative Extension: Madison, WI, USA, 2003; pp. 18–24.
27. Wolf, A.; Watson, M.; Wolf, N. Digestion and dissolution methods for P, K, Ca, Mg, and trace elements. In *Recommended Methods of Manure Analysis*; Publication No. A3769; University of Wisconsin Cooperative Extension: Madison, WI, USA, 2003; pp. 30–38.
28. Ayadi, F.Y.; Cortus, E.L.; Spiehs, M.J.; Miller, D.N.; Dijira, G.D. Ammonia and greenhouse gas concentration at surfaces of simulated beef cattle bedded manure packs. *Trans. ASABE* **2015**, *58*, 783–795.
29. Miller, D.N.; Varel, V.H. In vitro study of the biochemical origin and production limits of odorous compounds in cattle feedlots. *J. Anim. Sci.* **2001**, *79*, 2949–2956. [CrossRef] [PubMed]
30. Miller, D.N.; Varel, V.H. An in vitro study of manure composition on the biochemical origins, composition, and accumulation of odorous compounds in cattle feedlots. *J. Anim. Sci.* **2002**, *80*, 2214–2222. [CrossRef] [PubMed]
31. Dai, X.R.; Blanes-Vidal, V. Emissions of ammonia, carbon dioxide, and hydrogen sulfide from swine wastewater during and after acidification treatment: Effect of pH, mixing, and aeration. *J. Environ. Manag.* **2013**, *115*, 147–154. [CrossRef] [PubMed]
32. Hellman, B.; Zelles, L.; Palojarvi, A.; Bai, Q. Emission of climate-relevant trace gases and succession of microbial communities during open-windrow composting. *Appl. Environ. Microbiol.* **1997**, *63*, 1011–1018.
33. Mackie, R.I.; Stroot, P.G.; Varel, V.H. Biochemical identification and biological origin of key odor components in livestock waste. *J. Anim. Sci.* **1998**, *76*, 1331–1342. [CrossRef]
34. Batstone, D.J.; Keller, J.; Angelidaki, I.; Kalyuzhnyi, S.V.; Pavlostathis, S.G.; Rozzi, A.; Sanders, W.T.M.; Siegrist, H.; Vavilin, V.A. The IWA anaertobic digestion model no. 1 (ADM1). *Water Sci. Technol.* **2002**, *45*, 65–73. [CrossRef]
35. Batstone, D.J.; Keller, J. Industrial applications of the IWA anaerobic digestions model no. 1 (ADM1). *Water Sci. Technol.* **2003**, *47*, 199–206. [CrossRef]
36. Arikan, O. Effect of temperature on methane production from field-scale anaerobic digesters treating dairy manure. In *Waste to Worth: Spreading Science and Solutions*; Livestock Poultry Environmental Learning Center: Seattle, WA, USA, 2015.
37. Chandra, R.; Takeuchi, H.; Hasegawa, T. Methane production from lignocellulosic agricultural crop wastes; A review in context to second generation of biofuel production. *Renew. Sustain. Energy Rev.* **2012**, *16*, 1462–1476. [CrossRef]
38. Tauseef, S.M.; Premalatha, M.; Abbasi, T.; Abbasi, S.A. Methane capture from livestock manure. *J. Environ. Manag.* **2013**, *117*, 187–207. [CrossRef]
39. Environmental Protection Agency. Inventory of U.S. Greenhouse Gas Emissions and Sinks 1990–2015. 2017. Available online: https://www.c2es.org/content/internation-emissions (accessed on 18 June 2018).
40. Reis, M.A.M.; Almeida, J.S.; Lemos, P.C.; Carrondo, M.J.T. Effect of hydrogen sulfide on growth of sulfate reducing bacteria. *Biotechnol. Bioeng.* **1992**, *40*, 593–600. [CrossRef] [PubMed]

environments

MDPI

Article

Aeration to Improve Biogas Production by Recalcitrant Feedstock

John Loughrin * and Nanh Lovanh

United States Department of Agriculture, Agricultural Research Service, Food Animal Environmental Systems Research Unit, 2413 Nashville Road, Suite B5, Bowling Green, KY 42101, USA; nanh.lovanh@usda.gov

* Correspondence: john.loughrin@usda.gov; Tel.: +1-270-781-2260

Received: 19 March 2019; Accepted: 9 April 2019; Published: 11 April 2019

Abstract: Digestion of wastes to produce biogas is complicated by poor degradation of feedstocks. Research has shown that waste digestion can be enhanced by the addition of low levels of aeration without harming the microbes responsible for methane production. This research has been done at small scales and without provision to retain the aeration in the digestate. In this paper, low levels of aeration were provided to poultry litter slurry through a sub-surface manifold that retained air in the sludge. Digestate (133 L) was supplied 0, 200, 800, or 2000 mL/day air in 200 mL increments throughout the day via a manifold with a volume of 380 mL. Digesters were fed 400 g of poultry litter once weekly until day 84 and then 600 g thereafter. Aeration at 200 and 800 mL/day increased biogas production by 14 and 73% compared to anaerobic digestion while aeration at 2000 mL/day decreased biogas production by 19%. Biogas quality was similar in all digesters albeit carbon dioxide and methane were lowest in the 2000 mL/day treatment. Increasing feed to 600 g/week decreased gas production without affecting biogas quality. Degradation of wood disks placed within the digesters was enhanced by aeration.

Keywords: anaerobic digestion; bioenergy; biogas; carbon dioxide; methane; micro-aeration

---

1. Introduction

In 2012 over 8.5 billion broiler chickens were reared in the US [1]. Poultry are usually raised on bedding composed of absorbent materials termed "litter"; used poultry litter is usually applied to fields as fertilizer due to its high nitrogen and phosphorus content with or without prior composting. Its relatively high carbon content often depletes a considerable portion of the litter's nitrogen content during decomposition, though, reducing its fertilizer value [2]. In addition, the high phosphorus content of poultry litter is responsible for eutrophication of much of the nation's waters, especially in the southeast US where poultry rearing operations are especially concentrated [3].

Anaerobic digestion of animal wastes is an attractive option to controlling pollution caused by the large amounts of wastes generated in large animal rearing operations. Not only does it reduce the volume and strength of the waste, it also produces a crude natural gas (biogas) that can be used as fuel and thereby reduces the emission of methane, a potent greenhouse gas (GHG).

Animal waste, however, can often be a relatively poor feedstock for anaerobic digestion. This is because manures are often composed of complex carbohydrates (e.g., cellulose, pectin) and other substances such as lignin that are normally poorly degraded in anaerobic environments [4,5]. Poultry litter is a prime example of this, with high crude fiber content [2] due to the use of bedding materials such as wood chips, rice hulls, and straw, as well as high ash content resulting from decomposition of the litter and manure in the housing. Due to the high numbers of poultry reared, improved treatment technologies for the resulting waste are urgently needed to reduce pollution of the nation's waters. However, without means of improving yields of biogas from, and digestibility of, poultry litter, anaerobic digestion of poultry litter is unlikely to achieve widespread adoption.

Previous research has demonstrated that biogas yield and waste degradation can be improved by the addition of small amounts of air or oxygen to anaerobic digestates [6–8]. This has been termed "micro-aeration" wherein sufficient amounts of air are introduced to an anaerobic digestion to sustain the growth of organisms capable of degrading polymers such as cellulose that are normally persistent in anaerobic environments without harming the growth of the obligately anaerobic archaea or bacteria responsible for the production of methane or volatile fatty acids. For instance, Fu et al. [9] found that populations of oxygen-tolerant *Methanosarcina* and *Methanobacterium* were doubled under micro-aerobic conditions and, interestingly, that the abundances of strictly anaerobic *Clostridiales* were also increased under the same conditions.

Most of this research, however, has been performed upon a small scale and as a single batch fed experiment (e.g., [10–12]) without provision for retaining aeration below the surface of the digestate. This could present a problem in real world situations since much of the aeration could potentially be wasted by escaping the digestate and perhaps even dilute the biogas. This would be analogous to wastewater treatment in which aeration is typically inefficient, the majority of the air escaping to the atmosphere [13].

The aim of this research was to construct prototypes of anaerobic digesters that utilize a sub-surface manifold to retain aeration within the sludge layer of the digestate and more specifically within the manifold itself, where oxygen could be consumed without harmful effects on the microbial community responsible for biogas production. By retaining the aeration within the manifold, it was reasoned that the efficiency of micro-aeration would be increased.

The digesters were designed to simulate farm digesters treating waste fed on a continuing and regular basis (i.e., as would be employed in pit-recharge animal housing) over a prolonged period. In addition to ostensibly anaerobic conditions (discounting dissolved oxygen received during feeding), three levels of aeration over a ten-fold range were employed to gauge their effect on biogas yield and quality.

2. Materials and Methods

2.1. Digester Descriptions

Digesters were constructed from 208 L (55 gallon) blow-molded applicator tanks (US Plastic Inc., Lima, OH, USA). The side of each tank had a hole drilled into it to accommodate a 5.08 cm diameter polyvinyl chloride (PVC) pipe fitted with a manually operated ball valve that served as a waste inlet. This pipe extended into the tank below the surface of the digestate liquid. Float level switches (Omega Engineering Inc., Norwalk, CT, USA) were installed in the side of the tanks to maintain a digestate volume of 133 L. The float level switch was used to activate an electrical relay (American Zettler, Inc., Aliso Viejo, CA, USA) routing power to a 1.27 cm full port solenoid-actuated 120-VAC PVC ball valve (Valworx, Inc., Cornelius, NC, USA) installed on a 1.27 cm diameter PVC pipe that served as the waste outlet.

The waste outlet pipe had a hole drilled into it that accommodated a 0.3175 cm diameter line that led into the interior of the tank and provided aeration to a 2.54 cm diameter manifold constructed of PVC pipe installed in the bottom of the digester tank. The manifold had an "H" configuration with the long arms (0.6 m) of the manifold extending the length of the tank and the short arms (0.3 m) extending the width of the tank with a resulting volume of approximately 380 mL. The manifold had end caps installed on the end of its long arms and the bottom of the manifold had a series of 0.3175 cm diameter holes drilled in it allowing communication to the sludge layer of the digestate. Aeration was supplied to the subsurface manifold in 200 mL increments over 15 min intervals 0, 1, 4, or 10 times daily. The air was supplied through a 15 cm tall flow meter supplied by a diaphragm air pump and a 12-volt DC solenoid-actuated gas valve (Spartan Scientific, Boardman, OH, USA) controlled by a rotary timer. The aeration periods were spaced at equal intervals throughout the day.

The cap of each tank was adapted to accommodate a 3-way luer valve and 0.635 cm tubing that served as a gas outlet and sampling port. The tubing was connected to a Wet Tip Flow Meter® (wettipgasmeter.com) by one arm of a 3-way luer valve fitting. The other arm of the fitting accommodated a syringe for taking samples for gas analysis. The side of the tank had an additional 0.635 cm diameter port installed for liquid analysis. All pipe connections to the tanks were made with Uniseal® pipe to tank fittings (US Plastic, Inc.). A schematic of a micro-aerated digester is presented as Figure 1.

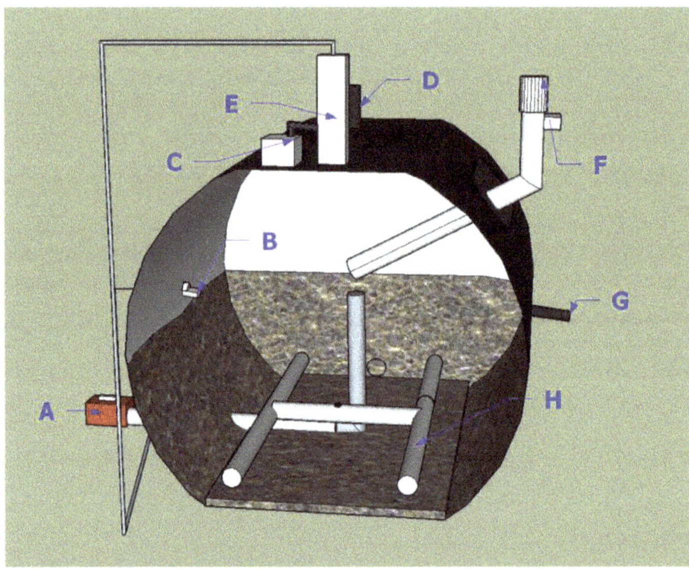

Figure 1. Cutaway schematic of a micro-aerated digester: (A) Waste outlet valve; (B) Float level switch; (C) Air pump; (D) Gas sampling port; (E) Flow meter for aeration; (F) Feed inlet; (G) Wastewater sampling port; (H) Aeration "H"-manifold.

2.2. Digester Operation

The digesters were kept in a greenhouse maintained at 26.7 °C. The digesters were 'seeded' with 20 L of liquid obtained from an anaerobic digester located on a commercial broiler operation in Kentucky. Digesters were then fed 400 g of poultry litter in 4 L deionized water once weekly until day 84 and from then on fed 600 g once weekly. The poultry litter averaged 40.4 ± 6.5% moisture with a volatile solids (VS) and ash content of 74.7 ± 3.6% and 25.3 ± 3.6%, respectively, on a dry weight basis. Gas production was measured daily during the workweek and averaged over the weekends. Gas quality, dissolved gas content, and wastewater quality were measured weekly.

At the beginning of the experiment, each tank had seven disks of tulip poplar wood (*Liriodendron tulipifera* L.) placed within them with a diameter and thickness of 5.72 and 1.91 cm, respectively. The disks were dried at 105 °C for three days and weighed prior to placing them in the digesters. They had an average weight of 55.9 ± 6.1 g. At the end of the experiment, the disks were removed, cleaned and dried at 105 °C for three days prior to reweighing.

2.3. Analyses

All gas and wastewater analyses were performed on samples obtained from the digesters immediately prior to weekly feedings of poultry litter. Wastewater and solids analyses were performed per standard methods [14]. Dry weights of poultry litter were determined by heating the litter for 24 h at 105 °C while ash content was determined after drying for 24 h at 550 °C.

Dissolved GHG, bicarbonate, and carbonate (HCO_3^-, CO_3^{2-}) were measured by collecting 0.5 mL water samples using a syringe with an 18-gauge needle. The sample was then injected into a 20 mL vial filled with 9.5 mL of 0.1 N HCl and fitted with a rubber septum.

Total CO_2 (solvated CO_2, HCO_3^-, and CO_3^{2-}) concentrations were determined by:

$$\Sigma\, CO_2,\ mM = 20 * \left[\frac{(0.8 * Conc + Conc)}{1000\ \mu g\ mg^{-1}} * \frac{1}{44.01\ mg\ mmol^{-1}}\right] \tag{1}$$

where 20 is a multiplication constant accounting for 0.5 mL injections onto the gas chromatograph, 0.8 is the dimensionless Henry's constant (KH) for CO_2, and Conc is the CO_2 concentration in the gas vial in $\mu g\ L^{-1}$. The sum of the HCO_3^- and CO_3^{2-} concentrations were determined by a use of the Henderson–Hasselbalch equation [15]:

$$\Sigma HCO_3^-, CO_3^{2-}, mM = \frac{[Total\ CO_{2,mM}]}{1 + 10^{(pH-6.35)}} * 10^{(pH-6.35)} \tag{2}$$

where pH is the pH of the solution and 6.35 equals the pk_{a1} for $H_2O + CO_2 \rightleftharpoons HCO^{3-} + OH^-$. Carbonate was calculated using the formula:

$$CO_3^{2-},\ mM = \frac{[Total\ CO_2, mM]}{1 + 10^{(pH-10.33)}} * 10^{(pH-10.33)} \tag{3}$$

with the variables the same as in Equation (2) and substituting a pk_{a2} of 10.33 for $HCO_3^- \rightleftharpoons CO_3^{2-} + H^+$. Bicarbonate concentrations were calculated by subtracting CO_3^{2-} concentrations from those calculated in Equation (3) from the concentrations calculated in Equation (2). Solvated CO_2 (sCO_2) concentrations were calculated by subtracting the values calculated from Equation (2) from those calculated in Equations (1) and (3).

Aqueous CH_4 and N_2O concentrations in the wastewater were calculated by Equation (1) using dimensionless Henry's constants of 27.02 and 1.1, respectively, and molar masses of 16.04 mg $mmol^{-1}$ and 44.01 mg $mmol^{-1}$, respectively [16,17].

Gas chromatographic analyses were performed as previously described [18] and statistical analyses were performed using SAS version 9.3 (SAS Institute, Cary, NC, USA).

3. Results and Discussion

3.1. Gas Production

Averaged for all four treatments, the digesters produced 0.4 m^3 of biogas per kg of volatile solids which compares favorably to a previously reported potential of 0.8 m^3 biogas per kg of volatile solids for poultry litter [15]. Aeration at 200 or 800 mL/day improved gas production over that of strictly anaerobic digestion by 14 and 73%, respectively (Figure 2). Conversely, aeration at 2000 mL/day decreased gas production by 19% as compared to the strictly anaerobic digester.

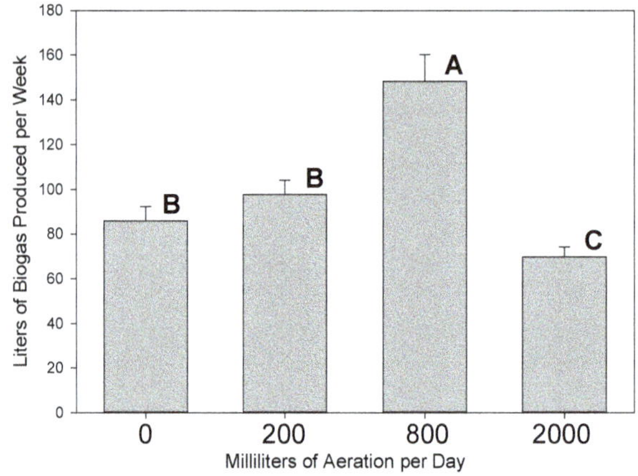

Figure 2. Average weekly gas production by digesters from week three onwards. Bars represent the mean of 18 determinations ± standard error of the mean. Bars labeled by the same letter are not significantly different by a Duncan's multiple range test.

From day 84 onwards, digesters were fed 600 g of poultry litter once weekly rather than the 400 g that had been fed previously. This had a negative effect on gas production in all four digesters. From day 59 to 83, gas production in all four digesters averaged 19.4 L/day while from day 84 onwards gas production averaged only 14.6 L/day. The digester receiving 2000 mL/day of supplemental aeration suffered the greatest decline in gas production at 34.1% whereas all other treatments averaged losses of 22.6%. This indicated that, regardless of treatment, either the capacity of the digesters for waste treatment had been exceeded or the waste loading rate had been increased too quickly.

As stated, part of the rationale for introducing micro-aeration to the digesters via a manifold was that the air would at least in part be retained within the manifold, so it could be consumed more efficiently. Of course, the belief that the aeration was retained within the manifold rests upon the supposition that it diffused into the digestate and the O_2 was consumed before the next aeration period. With the 2000 mL/day treatment, this meant air was supplied once every 2.4 h. Whether this time was enough to consume O_2 between aeration periods is unknown.

Two hundred, 400, and 800 mL of aeration per day introduced 1.5, 6.0, and 15 mL air per L digestate daily and roughly 0.45, 1.8, and 4.5 mg O_2 daily per L of digestate. In previous research, Xu et al. [12] obtained enhanced CH_4 production from a micro-aerated upflow anaerobic sludge blanket digester when aeration was supplied at a rate of 265 L air per kg VS and reduced CH_4 production at an aeration rate of 399 L air per kg VS. Here, we found significantly enhanced biogas production at 800 mL/day aeration which from day 0–84 of the experiment corresponded to 46.4 L air per kg VS on a weekly basis and decreased biogas production at 2000 mL of aeration per day which over this same time corresponded to 116 L air per kg VS. These numbers, however, are estimates since they do not account for accumulation of undigested feed during the experiment. It is problematic to compare our results to results of Xu et al. [12] since they used a different digester design and feedstock. It is possible, nevertheless, that we achieved enhanced production of biogas at lower aeration rates due to retention of the aeration within the manifold.

Although aeration at 800 mL/day significantly increased gas production, and aeration at 2000 mL/day significantly decreased gas production compared to the anaerobic digester, biogas quality was similar in all digesters. Nevertheless, biogas CO_2 concentrations did decline with increasing aeration (Table 1). Similarly, there was little difference in CH_4 concentrations between the various digesters. Aeration at 200 mL/day increased CH_4 concentrations by 8.7% compared to the strictly anaerobic digester whereas aeration at 800 and 2000 mL/day decreased CH_4 concentrations by 1% and 6.3%, respectively.

Table 1. Biogas and wastewater quality characteristics of digesters [1].

	Milliliters of Aeration Per Day			
	0	200	800	2000
pH	6.99 (0.09) a	6.98 (0.09) a	6.94 (0.08) a	6.93 (0.08) a
		Biogas Concentration (μg/L)		
CO_2	755,000 (21,300) a	740,000 (17,700) a	718,000 (17,800) a	683,000 (16,500) a
CH_4	318,000 (14,300) a	331,000 (24,600) a	314,000 (17,300) a	298,000 (9650) a
		Wastewater Concentration (Millimolar)		
HCO_3^-	49.8 (6.0) b	53.8 (6.6) a	54.5 (7.2) a	57.2 (7.1) a
sCO_2	9.1 (0.9) a	9.7 (0.8) a	10.5 (0.7) a	12.4 (1.0) a
sCH_4	21.1 (0.5) a	21.7 (0.5) a	23.5 (0.7) a	26.2 (0.5) a
		Wastewater Concentration (mg/L)		
Chemical oxygen demand	3240 (194) a	2720 (120) b	2760 (98) b	2420 (96) b
Total suspended solids	287 (29) a	276 (24) a	241 (27) a	262 (28) a
NH_4^+	153 (83) a	162 (65) a	150 (65) a	154 (68) a

[1] Values represent the mean of 18 determinations (standard error of the mean). Within rows, means followed by the same letter are not significantly different by a Duncan's multiple range test.

As stated, from day 84 onwards, digesters were fed 600 g once weekly rather than the previous 400 g. While this had negative consequences on gas production, gas quality was not similarly affected. Carbon dioxide and CH_4 biogas concentrations were quite similar before and after increasing to 600 g per week feedings.

Previous research has shown increases in bicarbonate buffering during micro-aeration of digestates [19] and we also noted increases in bicarbonate concentrations with increasing levels of aeration (Table 1). This likely explains the lower concentrations of CO_2 in digesters receiving aeration when compared to the anaerobic digester. To our knowledge, no previous research has conducted measurements of soluble CO_2 so the question of whether micro-aeration also affects sCO_2 concentrations has likely not been previously addressed. Here, we noted that nominally sCO_2 concentrations increased with the higher levels of aeration. We refer to CO_2 as nominally soluble because we feel that a considerable portion of the CO_2 is not solvated but rather in the form of bubbles, either free or attached to solids and other surfaces within the digesters. This supposition is supported by the fact that whereas bicarbonate concentrations increased smoothly throughout the experiment in all digesters (Figure 3), in the digester receiving 2000 mL of air per day considerable surges in sCO_2 were noted at weeks 11, 15, and 17 onwards. These surges are best envisioned as being due to bubbles of CO_2 rather than as being caused by increases in the concentration of solvated gas. Still, in general, sCO_2 concentrations tended to fall as the pH and HCO_3^- buffering of the digesters increased. In contrast to the behavior of sCO_2, sCH_4 concentrations reached their maximum concentrations by week four of the experiment and did not increase thereafter (data not shown).

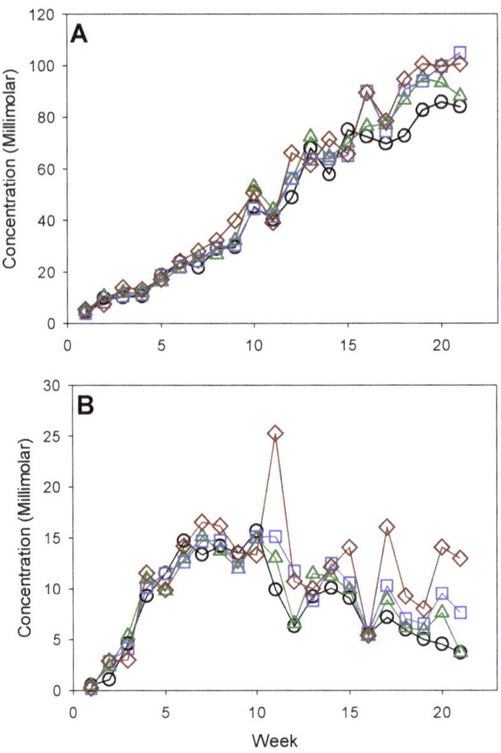

Figure 3. Bicarbonate (**A**) and soluble CO_2 (**B**) concentrations in digesters. Black line, circles: 0 mL supplemental aeration; green line, triangles: 200 mL aeration; blue lines, squares: 400 mL aeration; red lines, diamonds: 2000 mL aeration.

We did not measure H_2S or HS^- in our experiments. However, research has demonstrated a decrease in biogas H_2S concentrations due to micro-aeration [19]. This has been ascribed to the removal of HS^- and S^{2-} ions as elemental sulfur at O_2 tensions of less than 0.1 mg/L or sulfide oxidation to sulfate at higher O_2 tensions [20]. It is important to note that H_2S is often described as being toxic to methanogens and hence potentially reducing biogas yields [7,20,21]. It is more likely, however, that sulfate reducing bacteria (SRB) and archaea (SRA) outcompete methanogens for H_2 rather than H_2S exhibiting toxicity to methanogens per se [22]. In a micro-aerated digester, it is possible for sulfate ions to be formed which could encourage the activity of SRB and SRA. Although we did not note a corresponding increase in digestate pH as would likely be expected with enhanced HCO_3^- buffering, normally increased buffering would be expected to raise pH and thereby decrease biogas H_2S concentrations given that the pk_a for H_2S is 6.9.

3.2. Waste Degradation

Gas production was increased slightly in the digesters receiving 200 and especially 800 mL/day of aeration compared to the anaerobic digester. Conversely, at 2000 mL/day, gas production was likely decreased by oxygen inhibition of the microbial consortia responsible for methane production. It would be expected, therefore, that waste degradation would be enhanced by all treatments as compared to the anaerobic control. We did find that chemical oxygen demand (COD) was significantly reduced by all treatments as compared to the anaerobic digester (Table 1). While no significant differences in total suspended solids (TSS) were seen in any treatment as compared to the anaerobic digester, TSS were lower in all the aerated digesters.

While biogas production was inhibited by increasing weekly feedings to 600 g, increases in COD and TSS concentrations were not as great as might be expected. COD concentration in the anaerobic, 200, and 2000 mL/day averaged 3320, 2670, and 2570 mg/L from day 59 to 84, respectively, and 3760, 3040, and 2570 mg/L onwards. COD concentration was lowered in the 800 mL/day treatment as compared to the anaerobic digester, averaging 3010 mg/L from day 59 to 84 and 2910 mg/L afterwards. TSS averaged 329, 284, 368, and 359 mg/L from day 59 to 84 in the anaerobic, 200, 800, and 2000 mL/day aeration treatments, respectively, and 355, 341, 239, and 284 mg/L afterwards.

Ammonium concentrations were not affected by micro-aeration treatment (Table 1). No significant concentrations of either nitrate or nitrite were found in any of the wastewater samples (data not shown), nor was dissolved N_2O. This shows that even in the digester receiving 2000 mL/day aeration, conditions did not support any significant nitrification/denitrification.

As stated, a rationale for this study was that much agricultural waste is to a large extent composed of substances (e.g., wood) that is recalcitrant to degradation in anaerobic environments. The producer from whom the poultry litter was obtained indicated that the bedding material consisted of either an unspecified *Pinus* species or tulip poplar (*L. tulipifera*) wood chips depending upon price and availability. Therefore, to test whether the supplemental low-level aeration would facilitate breakdown of the bedding material, we added wood disks cut from tulip poplar boards to the digestate at the beginning of the experiment. Figure 4 represents wood disk weights at the beginning of the experiment and after incubation in the digesters for 148 days.

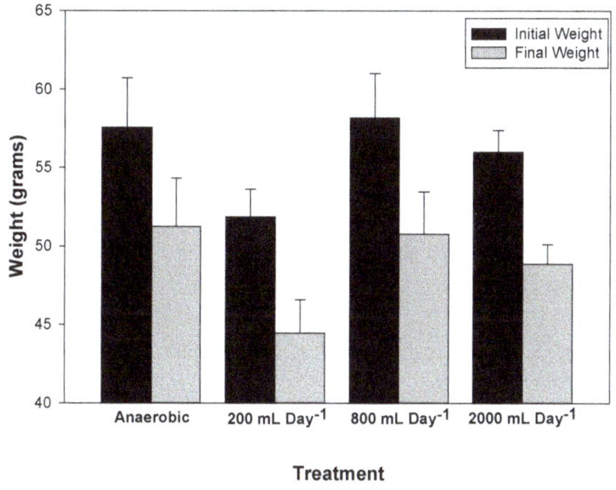

Figure 4. Weights of wood disks at beginning and end of experiment. Data represent the mean of seven determinations ± standard error of the mean.

Although significant weight loss occurred in the experiment for all treatments ($p < 0.0001$, *t*-test of $\log_{10}$ transformed data), there were no significant differences seen in weight loss among treatments. Nevertheless, wood disks placed in the strictly anaerobic tanks lost the least amount, averaging 6.3 g on a dry weight basis whereas all other treatments lost over 7 g dry weight. It has been shown that micro-aeration of wastes increases the hydrolysis of polymers such as cellulose as compared to strictly anaerobic conditions [6–8].

Fungal growth was noted on the wood disks in digesters receiving micro-aeration but most notably in the digester receiving 2000 mL/day air. No fungal growth was noticed on wood disks in the strictly anaerobic digester. It has been reported that biogas production can be improved in the presence of white rot fungi and anaerobic fungi such as *Neocallimastigomycota* [23] due to their ability to degrade complex polymers, and that the cellulose degradation activity of anaerobic fungi is increased in the presence of methanogens [24]. Analyses are planned to identify this fungus as well as determine how bacterial and other microbial populations may have been affected by micro-aeration.

4. Conclusions

At least three outcomes are desired in the micro-aeration of anaerobic wastes: that the production of biogas is enhanced, that waste degradation is enhanced, and that bicarbonate buffering is enhanced so that the stability of the digestion process is improved. Data from this experiment shows that all these goals can be achieved by micro-aeration. It also shows that considerable manipulation of aeration rates may be needed to achieve optimal results. It also shows that if waste loading rates are increased during digestion, it should be done gradually to maintain ideal digestion rates and avoid overloading digesters.

Author Contributions: Investigation, J.L. and N.L.; Methodology, J.L. and N.L.; Project administration, J.L.

Funding: This research received no external funding and was conducted as part of USDA-ARS National Program 212: Soil and Water, Developing Safe, Efficient and Environmentally Sound Management Practices for the Use of Animal Manure.

Acknowledgments: The authors thank Stacy Antle, Mike Bryant, (USDA-ARS), and Zachary Berry (Department of Chemistry, Western Kentucky University) for technical assistance. The use of trade, firm, or corporation names in this web site is for the information and convenience of the reader. Such use does not constitute an official endorsement or approval by the United States Department of Agriculture or the Agricultural Research Service of any product or service to the exclusion of others that may be suitable.

Conflicts of Interest: The authors declare no conflict of interest.

References

1. National Agricultural Statistics Service. Available online: https://www.usda.gov/sites/default/files/documents/nass-poultry-stats-factsheet.pdf. (accessed on 8 March 2019).
2. Stephenson, A.H.; McCaskey, T.A.; Ruffin, B.G. A survey of broiler litter composition and potential value as a nutrient resource. *Biol. Wastes* **1990**, *34*, 1–9. [CrossRef]
3. Kibet, L.C.; Allen, A.; Kleinman, P.J.; Feyereisen, G.W.; Church, C. Phosphorus runoff losses from subsurface-applied poultry litter on coastal plain soils. *J. Environ. Qual.* **2011**, *40*, 412–420. [CrossRef]
4. Kirk, T.K.; Farrell, R.L. Enzymatic "combustion": The microbial degradation of lignin. *Ann. Rev. Microbiol.* **1987**, *41*, 465–505. [CrossRef] [PubMed]
5. Leschine, S.B. Cellulose degradation in anaerobic environments. *Annu. Rev. Microbiol.* **1995**, *49*, 399–426. [CrossRef] [PubMed]
6. Díaz, I.; Donoso-Bravo, A.; Fdz-Polanco, M. Effect of microaerobic conditions on the degradation kinetics of cellulose. *Bioresour. Technol.* **2011**, *102*, 10139–10142. [CrossRef] [PubMed]
7. Tsapekos, P.; Kougias, P.G.; Vasileiou, S.A.; Lyberatos, G.; Angelelidaki, I. Effect of micro-aeration and inoculum type on the biodegradation of lignocellulosic substrate. *Bioresour. Technol.* **2017**, *225*, 246–253. [CrossRef]
8. Sawatdeenarunat, C.; Sung, S.; Khanal, S.K. Enhanced volatile fatty acids production during anaerobic digestion of lignocellulosic biomass via micro-oxygenation. *Bioresour. Technol.* **2017**, *237*, 139–145. [CrossRef] [PubMed]
9. Fu, S.-F.; Wang, F.; Shi, X.-S.; Guo, R.-B. Impacts of microaeration on the anaerobic digestion of corn straw and the microbial community structure. *Chem. Eng. J.* **2016**, *287*, 523–528. [CrossRef]
10. Zhu, M.; Lü, F.; Hao, L.-P.; He, P.-J.; Shao, L.-M. Regulating the hydrolysis of organic wastes by micro-aeration and effluent recirculation. *Waste Manag.* **2009**, *29*, 2042–2050. [CrossRef]
11. Lim, J.W.; Wang, J.-Y. Enhanced hydrolysis and methane yield by applying microaeration pretreatment to the anaerobic co-digestion of brown water and food waste. *Waste Manag.* **2013**, *33*, 813–819. [CrossRef]
12. Xu, S.; Selvan, A.; Wong, J.W.C. Optimization of micro-aeration intensity in acidogenic reactor of a two-phase anaerobic digester treating food waste. *Waste Manag.* **2014**, *34*, 363–369. [CrossRef]
13. Malovanyy, M.; Shandrovych, V.; Malovanyy, A.; Polyuzhyn, I. Comparative analysis of the effectiveness of regulation of aeration depending on the quantitative characteristics of treated sewage water. *J. Chem.* **2016**, *2016*, 6874806. [CrossRef]
14. American Public Health Association (APHA). *Standard Methods for Examination of Water and Wastewater*, 20th ed.; APHA; AWWA; WPCF: Washington, DC, USA, 1998.

15. Po, H.N.; Senozan, N.M. The Henderson-Hasselbalch equation: Its history and limitations. *J. Chem. Educ.* **2001**, *78*, 1499–1503. [CrossRef]
16. Howard, P.H.; Meylan, W.H. *Handbook of Physical Properties of Organic Chemicals*, 1st ed.; Chemical Rubber Corporation Press: Boca Raton, FL, USA, 1997.
17. Jacinthe, P.A.; Groffman, P.M. Silicone rubber sampler to measure dissolved gases in saturated soils and waters. *Soil Biol. Biochem.* **2001**, *33*, 907–912. [CrossRef]
18. Loughrin, J.H.; Cook, K.L.; Lovanh, N. Recirculating swine waste through a silicone membrane in an aerobic chamber improves biogas quality and wastewater malodors. *Trans. ASABE* **2014**, *55*, 1929–1937. [CrossRef]
19. Nuguyen, P.H.L.; Kuruparan, P.; Visvanathan, C. Anaerobic digestion of municipal solid waste as a treatment prior to landfill. *Bioresour. Technol.* **2007**, *98*, 380–387. [CrossRef]
20. Krayzelova, L.; Bartacek, J.; Kolesarova, N.; Jenicek, P. Microaeration for hydrogen sulfide removal in UASB reactor. *Bioresour. Technol.* **2014**, *172*, 297–302. [CrossRef]
21. Zhou, W.; Imai, T.; Ukita, M.; Li, F.; Yuasa, A. Effect of limited aeration on the anaerobic treatment of evaporator condensate from a sulfite paper mill. *Chemosphere* **2007**, *66*, 924–929. [CrossRef]
22. Kristjansson, J.K.; Sch6nheit, P.; Thauer, R.K. Different K_s values for hydrogen of methanogenic bacteria and sulfate reducing bacteria: An explanation for the apparent inhibition of methanogenesis by sulfate. *Arch. Microbiol.* **1982**, *131*, 278–282. [CrossRef]
23. Da Silva, R.R.; Pesezzi, R.; Souto, T.B. Exploring the bioprospecting and biotechnological potential of white-rot and anaerobic Neocallimastigomycota fungi: Peptidases, esterases, and lignocellulolytic enzymes. *Appl. Microbiol. Biotechnol.* **2017**, *101*, 3089–3101. [CrossRef]
24. Marvin-Sikkema, F.D.; Richardson, A.J.; Stewart, C.S.; Gorrschal, J.C.; Prins, R.A. Influence of Hydrogen-consuming bacteria on cellulose degradation by anaerobic fungi. *Appl. Environ. Microbiol.* **1990**, *56*, 3793–3797.

environments

MDPI

Article

Production of Ethanol from Livestock, Agricultural, and Forest Residuals: An Economic Feasibility Study

Kyoung S Ro [1,*], Mark A Dietenberger [2], Judy A Libra [3], Richard Proeschel [4], Hasan K. Atiyeh [5], Kamalakanta Sahoo [6] and Wonkeun J Park [7]

1 USDA-ARS, Coastal Plains Soil, Water & Plant Research Center, Florence, SC 29501, USA
2 USDA-FS, Forest Product Laboratory, Madison, WI 53726, USA
3 Leibniz Institute for Agricultural Engineering and Bioeconomy, 14469 Potsdam-Bornim, Germany
4 Proe Power Systems, LLC., Medina, OH 44256, USA
5 Department of Biosystems and Agricultural Engineering, Oklahoma State University, Stillwater, OK 74078, USA
6 USDA-FS, Forest Product Laboratory, Madison, University of Wisconsin-Madison, Madison, WI 53726, USA
7 Clemson University, Pee Dee Research Education Center, Florence, SC 29501, USA
* Correspondence: kyoung.ro@usda.gov; Tel.: +1-843-669-5203

Received: 12 July 2019; Accepted: 15 August 2019; Published: 17 August 2019

Abstract: In this study, the economic feasibility of producing ethanol from gasification followed by syngas fermentation via commercially available technologies was theoretically evaluated using a set of selected livestock and agricultural and forest residuals ranging from low valued feedstocks (i.e., wood, wheat straw, wheat straws blended with dewatered swine manure, and corn stover) to high valued oilseed rape meal. A preliminary cost analysis of an integrated commercial system was made for two cases, a regional scale 50 million gallon (189,271 m^3) per year facility (MGY) and a co-op scale 1–2 MGY facility. The estimates for the minimum ethanol selling prices (MESP) depend heavily on the facility size and feedstock costs. For the 1–2 MGY (3785–7571 m^3/y) facility, the MESP ranged from $5.61–$7.39 per gallon ($1.48–$1.95 per liter) for the four low-value feedstocks. These high costs suggest that the co-op scale even for the low-value feedstocks may not be economically sustainable. However, the MESP for the 50 MGY facility were significantly lower and comparable to gasoline prices ($2.24–$2.96 per gallon or $0.59–$0.78 per liter) for these low-value feedstocks, clearly showing the benefits of scale-up on construction costs and MESP.

Keywords: swine manure; cover crops; wood; oilseed rape; syngas fermentation; gasification

1. Introduction

Sustainable agricultural biomass feedstock can be used to produce biofuel, bioenergy, biochemicals, and bioproducts via a variety of conversion pathways. Important sources of biomass feedstock for bioenergy identified by the 2016 Billion-Ton-Study and its update [1] include forest resources, energy crops, crop residues, and animal manures. While forest resources such as logging residues and the whole tree, and perennial energy crops (herbaceous: switchgrass, miscanthus, and energy cane, and woody crops: southern pine, poplar, willow, and eucalyptus) are often considered for bioenergy production, conversion processes for farm-based biomass such as crop residues and animal manures are less typical. Economically viable local conversion facilities to produce liquid biofuels would offer potential income to farmers. The development of such facilities is especially important since crop residues are one of the largest sustainable sources of feedstock in the United States. Corn stover and wheat straw are the two major crop residues. Further potential agricultural residual feedstocks are cover crops such as vetch, clover, and rye. These cover crops grown during fallow periods between major cash crops are used to prevent soil erosion. In addition to these traditional cover crops, the oil seed crop species *Brassica napus* represented by canola and rapeseed can also be used as a cover crop. As a cover crop, canola and rapeseed, cumulatively oilseed rape (OSR) is known for benefits such as preventing erosion, increasing soil organic matter, suppressing weeds and soilborne pests, alleviating subsoil compaction, and scavenging nutrients [2,3].

Animal manure is also considered to be one of the major available biomass waste resources [1]. In view of the change in animal production in the U.S. and worldwide toward more concentrated animal feeding operations, environmentally sound methods for the storage and disposal of this surplus manure are required [4]. Studies have shown that thermochemical conversion of surplus animal manures such as pyrolysis not only alleviates the storage and disposal problems, but it can also provide regionally available power for local farmers [4–7] and value-added byproducts that can be used to remove odors [8], fugitive gas such as ammonia [9,10], and as soil amendments to improve soil quality [11,12].

Gasification is another thermochemical technology which converts surplus manures, as well as other residues to produce synthesis gas (hereafter referred as syngas) composed mainly of CO and H_2. The syngas can be used to generate power via combustion or converted to liquid fuels via downstream catalytic syngas conversion [13] or fermentation processes. A number of commercial gasification systems capable of converting a variety of feedstocks for syngas production have been used [14–21]. The fermentation of the syngas in a following process step utilizes autotrophic microorganisms to convert CO, H_2, and CO_2 with flexible molar ratios into alcohols, organic acids, and other products [22–36]. Both monocultures and mixed cultures have been used to produce a variety of fermented products. Commercial facilities have been constructed, producing ethanol from municipal solid and industrial cellulosic wastes [14,15]. In addition to ethanol, research on competing uses for biomass continues and may bring forth alternative economically sustainable products, ranging from higher value niche chemicals [37,38] to lower cost bulk biofuels using chemical transformations (e.g., methanol, dimethyl ether, synthetic natural gas) [39].

With current gasification technology, low-valued agricultural residuals can be combined with the amenable woody feedstocks to increase the economic feasibility of ethanol production facilities. Small scale systems that draw farm-based and forest residues from surroundings with shorter biomass draw radius, i.e., shorter transport distance for hauling biomass to the plant, will reduce a major costing of the biomass supply [40,41]. Moreover, farm-based biomass can often include woody feedstocks. The relatively clean, high growth rate and the high heating values for some wood (particularly poplar) often motivate a creation of energy crop production alongside the agricultural production.

The overall goal of this study is to estimate the economic potential of producing ethanol from agricultural and forest residuals using commercially available technology for gasification-syngas fermentation. Specific objectives of this study were to determine (1) theoretical amounts of ethanol that can be produced from selected forest and agricultural residuals by operating coop- or regional-scale, integrated gasification, and syngas fermentation systems and (2) preliminary economic feasibility by conducting cost analyses of the integrated systems.

2. Materials and Methods

2.1. Feedstocks and Logistics

This study considered forest resources (logging residues), crop residues (corn stover and wheat straw), swine manure mixed with wheat straw, and OSR meal (a residual from oil processing), to produce liquid biofuel for two cases from small to large commercial systems. Case 1 was assumed to be a regional-scale size bioethanol facility (50 million gallon per year, 50 MGY or 189,271 m^3/y). Case 2 was assumed to have 1 to 2 MGY (3785–7571 m^3/y) ethanol production capacities as a co-op-scale facility. The plant capacity and biomass availability surrounding a biofuel plant determine the biomass draw radius or transport distance between feedstocks and the plant.

The estimated roadside costs of forest resources that include stumpage price and harvesting costs (assuming an integrated harvesting option) were taken from the 2016 Billion-Ton-Study [1]. Both stumpage price and harvesting cost varied widely between regions, which mainly depend on harvesting method, topography, and stand type. Usually, crop residues are harvested as bales, especially large rectangular bales that have lower harvesting, handling, storage, and transportation cost compared to round bales [42,43]. The farmgate price of crop residues included nutrient replacement cost due to crop residues removal and harvesting cost.

The delivered cost of biomass to a biomass conversion plant includes roadside or farmgate cost and transportation cost of biomass from forest roadside or farm gate to bioenergy/biofuel conversion site. Storage may be necessary due to the seasonal availability of biomass, especially crop residues [43]. Forest resources and manure may be available throughout the year and biomass storage may not be necessary. A detailed estimation of bales and woodchips were described in Sahoo and Mani [43] and Sahoo et al. [44] respectively. The conversion plant considered in this study can use wood chips directly and further grinding is not required. However, bales require additional grinding operation before use in the conversion plant. A detailed estimation of transport and handling cost for bales and woodchips were presented by Sahoo and Mani [42] and Sahoo et al. [45] respectively. Animal manure cost at the farmgate varies between 0 and 40 $/dry tonne [1]. Because swine manure solid contents are only about 2–5%, various dewatering techniques are used to dewater the manure and reduce the transportation cost. The dewatered swine manure can be available at the price of $22/tonne (25% solid content) (personal communication with a manure sludge management company in North Carolina). Although the cost for OSR meal is much higher than other feedstocks, it was selected for comparing its competing use as animal feed. Table 1 summarizes the logistics cost of various feedstocks considered in this study.

Table 1. Logistics costs of biomass for the small-scale portable biofuel plants.

Biomass Type		Stumpage Cost [a] ($/Dry Tonne) [1]	Harvesting Cost ($/Dry Tonne) [1]	Transportation Cost ($/Dry Tonne)		Storage Cost [d] ($/Dry Tonne) [43,44]	Grinding ($/Dry Tonne) [e] [42]	Average Total Feedstock Cost [f] ($/Dry Tonne)	
				20 [b] km	60 [c] km			20 [a] km	60 [c] km
Logging residues	Wood chips	1–5	15–20	6	12	8	-	34.5	40.5
Wheat Straw	Rectangular bale	9–28	44–49	4	10	6.0	10.5	85.5	91.5
Corn Stover	Rectangular bale	5–15	14–16	4	10	6.0	10.5	45.5	51.5
Swine Manure		70	0	3.5	8.0	-	-	73.5	78.0
Oilseed rape (OSR) meal		342	-	4	10	-	-	346.0	352.0
Wheat Straw (50%) + Swine manure (50%)		-	-	-	-	-	-	79.5	84.8

[a] A price paid to forest owners for logging residues. A price paid to farmers for the crop residues to recover the cost of adding removed nutrients due to the removal of crop residues [1]. Swine manure delivered cost was provided by the local vendors $22/tonne (25% solid content) includes transportation (personal communication) and this study assumed the cost $17.5/tonne (excluding transportation cost) of dewatered swine manure and estimated the transportation cost for the specific transport distances of 20 km ($3.5/dry tonne) and 60 km ($8/tonne) for small and large scale plants respectively. Oilseed rape (OSR) meal cost was taken from USDA year book ($285/US tonne, 8% moisture content [46]. [b] Assumed biomass draw radius of 20 km for smaller size plant (1–2 million gallon per year (MGY) ethanol plant capacity). Transport cost was estimated based on [42,45,47]. [c] Assumed biomass draw radius of 60 km for large size plant (50 MGY ethanol plant capacity). Transport cost was estimated based on [42,45,47]. [d] Wood chips are assumed to be stored for 1 month as they are available throughout the year. Crop residues are seasonal and thus a six-month storage of bales was assumed [43,44]. Storage was not considered for Swine manure and OSR meal. [e] Debaling and grinding cost for agricultural wastes [42]. Grinding or chipping is part of harvesting logging residues. [f] Average total feedstocks cost delivered to biorefinery includes average values for stumpage, harvest, transportation, storage and grinding.

2.2. Proximate and Ultimate Analyses of Feedstock

In order to estimate the syngas production rates and compositions using the Proe Power Systems' simulation model, proximate and ultimate properties of feedstock were needed. Triplicate proximate and ultimate properties of wheat straw samples (ASTM D3172 and 3176) were analyzed by Hazen Research Inc. (Golden, CO, USA). Mean values of proximate and ultimate properties reported in the literature were used for swine manure and OSR meal [48,49].

2.3. Syngas Production Using a Commercial Gasification System

In order to handle a wide variety of feedstocks, we chose to use a modular fixed-bed, crossflow gasification system (Proe Power 250 kWe, Proe Power Systems, LLC, Median, OH, USA) to generate syngas from various forest and agricultural residuals. Selecting the appropriate gasifier technology has several facets when using the farm-based biomass [16]. Due to the variation in the size of the biomass feedstock, many biomass gasification systems require attention to size reduction and/or feed flow uniformity, particularly in downdraft and fluidized bed gasifiers. The fixed bed gasifier was chosen because it was more tolerant of feedstock sizing, helping to reduce costs. Air was used in the gasifier instead of oxygen for the gasification reactions since oxygen separation units are usually not cost-effective at the smaller scales [17]. Likewise, the higher cost associated with the use of super-heated steam for carbon gasification [17,18] was also avoided at the smaller scales. However, there is a potential low-cost approach of using a commercial air heat exchanger to preheat the combustion air and lower the nitrogen content of the syngas. With the air heat exchanger, less oxygen is needed to reach the adiabatic temperature within the gasifier that then promote more CO and H_2 production and less CO_2 and H_2O byproducts via the water gas shift reactions; which has some mention as a concept in the literature [19,20]. A preheated combustion air concept in a downdraft gasifier design has already been used for gasification of cow manure [21], but the downdraft gasifier design may not be optimal for other types of farm feedstock.

The Proe Power Systems' simulation model for the crossflow gasification system component was used to determine the syngas composition and production rates. The model is a thermochemical equilibrium model based on the elemental composition of the feedstock and the assumption that equilibrium is achieved. The model can be used for both woody and non-woody biomass feedstocks. The composition was determined for gasification with two output temperatures of 649 Celsius and 850 Celsius. This simulation model predicts material flow rates and temperatures of various components of the gasification system. Figure 1 shows the basic schematic of the selected modular Proe Syngas Generator, the predicted material flow rates, and temperatures for gasifying wood fuel to generate syngas. The simulation model performs mass and energy balances on the air heater, gasifier, and dryer. Input values are biomass fuel properties, gasifier temperature, air temperature, desired syngas production flowrate, and the ratio of air flow split to the fuel dryer and gasifier. This model can be used to optimize the syngas production for a wide variety of biomass fuels by varying the input values. As a check we also used the gaseous equilibrium solver, known as StanJan, to verify the model accuracy on the major gases, at least to within 1% at 850 °C, and nil production of solid carbon at equilibrium. Predictions of the computer model are comparable with the literature and other models for the air blown gasifiers [16]. The composition is then considered "frozen" at that equilibrium value as it passes through the air heater and on to the fermentation process.

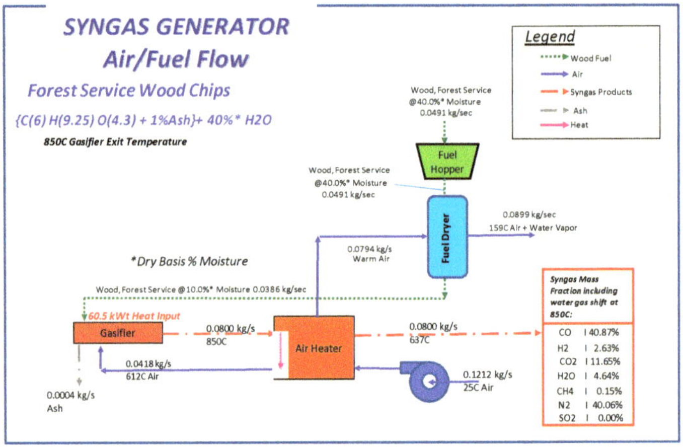

Figure 1. Calculated result of syngas generator for the wood chips feedstock.

2.4. Theoretical Ethanol Yields from Syngas Fermentation and Ethanol Production Rates

Syngas-utilizing microorganisms fix carbon via the acetyl-CoA pathway, also known as Wood–Ljungdahl pathway [50], and its derivative [51,52]. Acetic acid and ethanol are produced from CO, CO_2, and H_2 according to the following possible reactions, which are dependent on syngas composition [23]:

$$2\,CO + 2\,H_2 \rightarrow CH_3COOH \tag{1}$$

$$4\,H_2 + 2\,CO_2 \rightarrow CH_3COOH + 2\,H_2O \tag{2}$$

$$4\,CO + 2\,H_2O \rightarrow CH_3COOH + 2\,CO_2 \tag{3}$$

$$CO + 3\,H_2 + CO_2 \rightarrow CH_3COOH + H_2O \tag{4}$$

$$6\,CO + 3\,H_2O \rightarrow CH_3CH_2OH + 4\,CO_2 \tag{5}$$

$$3\,CO + 3\,H_2 \rightarrow CH_3CH_2OH + CO_2 \tag{6}$$

$$4\,CO + 2\,H_2 + H_2O \rightarrow CH_3CH_2OH + 2\,CO_2 \tag{7}$$

$$CO + 5\,H_2 + CO_2 \rightarrow CH_3CH_2OH + 2\,H_2O \tag{8}$$

In Equations (1)–(4), 1 mole of acetic acid is produced from 4 moles of reductants. For ethanol production Equations (5)–(8), 1 mole of ethanol requires 6 moles of reductants. The reductants in syngas fermentation come from either H_2, CO or both. In this study, the theoretical ethanol yield was estimated based on the assumption that all the reductants (i.e., 6 moles CO + H_2) available in the syngas were converted to one mole of ethanol [23]. This assumption was combined with the syngas yield from each feedstock to calculate theoretical ethanol yields, which are reported in the results section. In practice, though, not all the CO and H_2 is converted to ethanol by fermentative microorganisms, nor is all ethanol recovered from the system. In this study, we assumed 90% conversion of CO and H_2 to ethanol, and 90% ethanol recovery efficiency (or yield) of ethanol from the fermentation system to estimate the actual ethanol yields.

Since the fermentation process benefits from steady state conditions in the system, the feed rates of the five feedstocks to the gasifier were adjusted to obtain syngas flow rates that would yield similar ethanol production rates (case 1) or were similar (case 2). For case 1, the regional facility was sized to produce 50 MGY. Syngas flows ranged between 30 to 50 kg/s (Table 5). For case 2, a 1–2 MGY commercial level facility for farm-based application at the coop level would need around 1 kg/s of the syngas. A factor of 1000/80 = 12.5 was used to obtain the respective flows for the coop facility.

2.5. Cash Flow Analyses

Cost estimation was based on an 8 million gallon per year, 80 MGY(30,283 m^3/y), commercial demonstration ethanol facility built in Florida in 2013 [14]. The Ineos Bio facility produces 8 MGY ethanol from municipal solid waste, that includes gasification, fermentation, and distillation. The construction cost of the facility was reported to be $130 million in the 8 MGY facility that employed 65 full-time employees. A summary of assumptions made for cost estimation in the present study is shown in Table 2. The feedstock cost for the base case (Ineos Bio) was assumed to be $70 per dry tonne of feedstock. However, for the estimated cases in the present study, cases 1 and 2, the feedstock costs were based on the estimation provided in Table 1. The case 1 was assumed to have a typical size bioethanol facility (50 MGY) while the case 2 was assumed to have 1 to 2 MGY ethanol production capacities as small co-op facilities.

The construction cost and employee numbers were scaled by feed capacity ratio to the power 0.7 for the case 1. For the case 2, the number of operation employees was assumed to be 8. Other assumptions used in estimation are shown in Table 2. The minimum ethanol selling price (MESP) per gallon was estimated by equating revenue with the sum of feedstock cost, operating costs (supply and labor), and capital return per year.

Table 2. Summary of assumptions for cost estimation of the two cases.

	Base [a]	Case 1 [b]	Case 2 [c]
Nameplate Ethanol (MGY)	8	50	1 to 2
Construction Cost (M$)	130	Varied [d]	Varied [d]
Assumed Yield (gal/tonne)	100	Varied [e]	Varied [f]
Feedstock Cost (per tonne)	$70	Varied [g]	Varied [g]
Feedstock Requirement (M tonne,yr)	0.080	Varied [e]	Varied [f]
Operations Employees	69	Varied [h]	8
Payroll Burden per Employee/year	$60,000	$60,000	$60,000
Payout Period (years)	10	10	10
Interest for Capital Return (%)	5%	5%	5%
Operating costs (% of Capital Cost/yr)	10%	10%	10%

[a] Based on published data for Ineos Bio 8 million gallons per year (MGY) facility [14]. [b] Case 1 is for 50 MGY based on typical ethanol biorefinery. [c] Case 2 is for 1 to 2 MGY for a small-scale co-operation. [d] Construction cost scaled by feed capacity ratio, $(Case/Base)^{0.7}$. [e] See Table 5 for values for each feedstock. [f] See Table 6 for values for each feedstock. [g] see Table 1 for values for each feedstock. [h] Employee number scaled by feed capacity ratio, $(Case/Base)^{0.7}$.

3. Results and Discussion

3.1. Feedstock Characteristics

This study considered five raw feedstocks and/or mixtures to the syngas unit: wood chips, two crop residues (corn stover and wheat straw), a blend of wheat straw and swine manure, and OSR meal in the two cases. The characteristics of the 5 feedstocks used in the feed of the simulation model are shown in Table 3. Their original moisture content (MC) varied from 8% to 40%. The ash content ranged from 1–12.3%. The highest ash content was due to the blending of wheat straw with dewatered swine manure, thereby increasing the value from 3.4% to 12.3%. The increase in ash content does not directly affect gasification reactions in the crossflow fixed-bed gasifier design but does result in a mass fraction that must be disposed of as a solid and therefore reduces the amount of dry fuel weight available for gasification, and is a factor included in the simulation.

Table 3. Actual Proximate and Ultimate values used in gasification simulation.

Parameters	Wood Chips	Wheat Straw (WS)	1:1 Blend of WS and SM [49]	OSR Meal (OSRM) [48]	Corn Stover
			Proximate—dry basis (db)		
Feed MC ($\%_{db}$)	40	40	40	8	10
Dried MC ($\%_{db}$)	10	10	10	2	2.5
Ash ($\%_{db}$)	1	3.4	12.3	7.3	4.9
			Ultimate—dry and ash free basis (dafb)		
C ($\%_{dafb}$)	48.0	48.4	53.7	50.7	49.5
H ($\%_{dafb}$)	6.2	6.4	6.9	6.8	6.1
O ($\%_{dafb}$)	45.8	44.1	35.7	34.7	43.7
N ($\%_{dafb}$)	0	1.0	3.2	6.9	0.68
S ($\%_{dafb}$)	0	0.1	0.5	0.9	0.02

3.2. *Syngas Yields and Production Rates*

An overview of the modeled material flow rates and syngas composition for the simulated syngas production at two synthesis gas outlet temperatures (649 °C and 850 °C) in a modular fixed-bed, crossflow gasification system is given in Table 4. The values are all referenced to syngas flow rate of 80 g/s which approximates that found in the Proe Power System's 250 kWe mobile facility, processing dried wood at a flow rate of 44 g/s and 649 °C. The simulation results for both gasification temperatures show the production of high amounts of CO and H_2, in approximately equal molar ratios (Table 4). The higher temperature output of 850 °C requires approximately 12% lower wood flow rates to produce the same syngas flow at 649 °C. Other feedstocks require 10% (wheat straw) to 8% (OSR meal) lower feed flows to produce the same syngas flow rate. Since the simulation model is set to maximize the heating value of the syngas, it is not surprising that there is a wide variation of dryer air flows, with the greatest flow variation (between 17 to 148 g/s for the 80 g/s syngas) in the drying of the feedstock to an appropriate level. The most challenging feedstock was the manure and wheat blend with relatively low heating value and high ash, and it still gave a reasonable production of CO and H_2 for the fermentation process. The ability to produce a reasonable CO and H_2 concentration is largely due to the regenerative heat process provided by the preheated combustion air, a feature often not available in other gasifier designs. Although the StanJan chemical equilibrium software has confirmed minimizing the methane and elimination of solid carbon, the challenge remains in the effectiveness and speed of the tar, char, and methane conversion process, which means a relatively long exposure of the producer gas to quite high temperatures, which is enhanced with preheated combustion air [18].

Our models assumed most of feedstock S became SO_2, as Xu et al., (2011) reported that COS and H_2S were far less than SO_2 in various gasifier outputs [53]. As for N in the feedstock, the gas equilibrium calculations showed N being primarily N_2 with miniscule amounts of NOx, HCN, and NH_3. These N gas species distributions can be highly variable with any particular gasification technology and cannot be reliably predicted with our models. Moreover, a thermochemical equilibrium model cannot predict potential interactions between feedstocks. Therefore, since some gasifiers are better than others to approach full conversion with varying requirements for cleaning syngas for downstream processes, experimentation in a pilot scale facility is needed to determine these concentrations and interactions.

Table 4. Mass flow, temperatures, and syngas composition of gasification simulation, all referenced to syngas flow rate of 80 g/s.

Gasification Parameters	Wood Chips		Wheat Straw		50 Wheat Straw/50 Manure		OSR Meal		Corn Stover	
	649 °C	850 °C	649 °C	850 °C	649 °C	850 °C	649 °C	850 °C	649 °C	850 °C
Feed Flow (g/s)	56	49.1	45.3	40.8	38.4	35.3	29.8	27.5	36.6	33.4
Dried Feed Flow (g/s)	44	38.6	35.6	32.1	30.2	27.7	28.2	26.0	34.1	31.1
Pumped Air Flow (g/s)	182	121.2	193.4	132	119.7	94.1	80.8	72.6	118.9	90.7
Dryer Air Flow (g/s)	145.6	79.4	147.9	83.1	66.5	38.8	26.9	16.8	71.4	40.3
Gasifier Air Flow (g/s)	36.4	41.8	45.5	48.9	53.2	55.4	53.9	55.8	47.6	50.4
Ash Flow (g/s)	0.4	0.4	1.1	1.0	3.4	3.1	2.0	1.9	1.6	1.5
Syngas/dried feed (g/g)	1.8	2.1	2.2	2.5	2.6	2.9	2.8	3.1	2.3	2.6
Syngas Temperature (°C)	649	850	649	850	649	850	649	850	649	850
Heated Air Temperature (°C)	443	612	445	615	485	665	460	626	450	621
Cooled Syngas Temperature (°C)	521	637	477	585	416	509	423	524	460	565
CO (% mass)	43.42	40.87	30.76	30.42	26.46	26.45	28.93	28.29	35.51	34.14
H_2 (% mass)	2.93	2.63	2.27	1.98	1.92	1.68	1.84	1.65	2.06	1.84
CO_2 (% mass)	14.47	11.65	18.15	14.39	15.48	12.57	11.97	10.15	13.46	11.45
H_2O (% mass)	2.69	4.64	3.7	5.8	3.09	4.95	2.01	3.67	2.15	3.83
CH_4 (% mass)	1.58	0.15	1.02	0.09	0.73	0.07	0.69	0.07	0.84	0.08
N_2 (% mass)	34.89	40.06	44.03	47.26	51.97	53.96	53.84	55.59	45.88	48.58
SO_2 (% mass)	0	0	0.06	0.06	0.35	0.32	0.63	0.58	0.09	0.08
CO (% moles)	32.11	30.89	24.52	24.63	21.84	22.1	23.96	23.7	28.78	28.05
H_2 (% moles)	30.14	27.58	25.17	22.27	22.01	19.52	21.21	19.23	23.17	21.03
CO_2 (% moles)	6.81	5.61	9.21	7.41	8.13	6.68	6.31	5.42	6.95	5.99
H_2O (% moles)	3.1	5.45	4.58	7.3	3.97	6.43	2.71	4.78	2.71	4.89
CH_4 (% moles)	2.05	0.2	1.42	0.13	1.06	0.1	1	0.1	1.19	0.12
N_2 (% moles)	25.79	30.27	35.08	38.24	42.87	45.05	44.58	46.56	37.17	39.9
SO_2 (% moles)	0	0	0.02	0.02	0.12	0.12	0.23	0.21	0.03	0.03

3.3. Cost Analyses

The preliminary cost analysis for the five feedstocks showed that the differences between the feedstocks and gasification conditions (e.g., costs, syngas yields) were dampened when further processes and the facility construction costs were considered. For example, although the syngas yields were higher at 850 °C than at 649 °C, ranging from 13% (corn stover) and 17% (wood chips), this translated into only slightly higher ethanol yields (2% to 5%, respectively) in the fermentation process (Tables 5 and 6). The highest ethanol yield was obtained from wood chips (133 gal per tonne biomass at 850 °C), while the yields for the four non-woody feedstocks were quite similar (117.8 to 120.2 gal per tonne). Looking at the construction costs, the higher costs are associated with the non-woody feedstocks in the 50 MGY facility, ranging from $384 to $424 million for wood chips and corn stover, respectively (Table 5). However, this is reversed for the cost of ethanol facilities between 1 and 2 MGY, varying from $26 million for the wheat straw/swine manure mixture to $36 million for wood chips (Table 6).

Comparison of the estimated MESP for the two cases shows that the MESP is very affected by the size of ethanol facility and cost of feedstocks (Tables 5 and 6 and Figure 2). However, transportation costs play a very small role. For a typical 50 MGY facility, the MESP ranged between $2.28 and $2.96 per gallon for the low-value feedstocks and $5.13 per gallon for the high value OSR meal. The lowest price ($2.28 per gallon) was obtained for wood chips with the lowest feedstock cost of $40.5 per dry tonne, while the MESP for the high valued OSR meal ($352 per dry tone) was more than double at $5.13 per gallon. Tripling the transportation distance only increases the MESP by 5 cents on average. This demonstrates that the cost for the feedstock plays a considerable role in determining the MESP for the large-scale facility. In contrast, the costs for construction and operation play a larger role in the MESP for smaller scale facilities. The MESP ranged between $5.61 and $9.49 per gallon (Figure 2). This clearly shows the effects of scale-up and feedstock cost on construction cost and MESP, since ethanol yield per dry tonne of feedstock was the same for both cases.

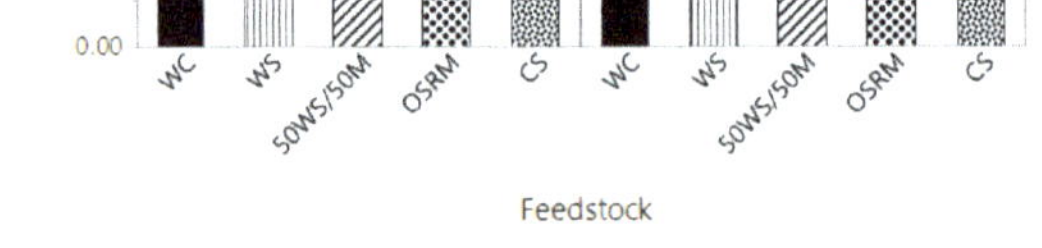

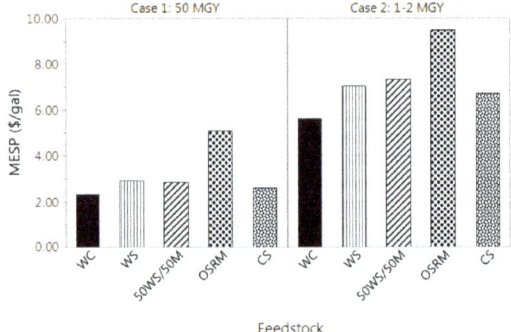

Figure 2. Minimum ethanol selling prices (MESP), for Case 1: 50 million gallons per year (MGY) facility (see Table 5; Table 1 for feedstock cost with transportation distance of 60 km) and Case 2: 1 to 2 MGY (see Table 6; Table 1 for feedstock cost with transportation distance of 20 km) for various feedstocks with gasification at 850 °C. WC: wood chips; WS: wheat straw; 50WS/50M: 50% wheat straw & 50% manure; OSRM: OSR meal; CS: corn stover.

Table 5. Case 1—Comparison of feedstock feed rate, syngas and ethanol yield and ethanol selling prices from gasification and syngas fermentation of 30–50 kg/s syngas from 5 feedstocks in a 50 million gallons per year (MGY) ethanol facility.

Feedstocks Parameters	Wood Chips		Wheat Straw		50 Wheat Straw/50 Manure		OSR Meal		Corn Stover	
	649 °C	850 °C	649 °C	850 °C	649 °C	850 °C	649 °C	850 °C	649 °C	850 °C
Syngas yield, kg/kg feedstock (db)	2.0	2.2	2.4	2.6	2.9	3.1	2.8	3.0	2.4	2.5
Syngas input (kg/s)	30.4	31.9	39.8	41.8	47.9	50.0	45.8	48.1	40.2	40.9
Feedstock feed, tonne (db)/d [a,b]	1311.3	1252.2	1433.1	1390.0	1425.7	1394.7	1412.4	1386.6	1445.5	1414.8
Annual demand, thousand tonne (db)/y [a,b]	393.39	375.66	429.92	417.01	427.72	418.41	423.73	415.97	433.65	424.45
Theoretical ethanol yield, gal/tonne (db) [a]	156.9	164.4	143.6	148.0	144.3	147.5	145.6	148.4	142.3	145.4
Ethanol selling price, $/gal [a]	1.98	1.92	2.47	2.41	2.42	2.38	4.23	4.16	2.21	2.17
Ethanol selling price, $/gal [b]	2.02	1.95	2.51	2.45	2.46	2.41	4.27	4.20	2.25	2.21
Ethanol yield, gal/tonne (db) [c]	127.1	133.1	116.3	119.9	116.9	119.5	118.0	120.2	115.3	117.8
Construction cost, $ million [a,b,c,d]	396.42	383.83	421.85	412.94	420.33	413.91	417.58	412.22	424.40	418.08
Operating cost, $ million [a,b,c,d]	50.67	49.08	53.87	52.75	53.68	52.87	53.33	52.66	54.19	53.39
Minimum ethanol selling price (MESP), $/gal [a,c]	2.31	2.24	2.91	2.84	2.84	2.79	5.08	5.00	2.58	2.54
Minimum ethanol selling price (MESP), $/gal [b,c]	2.36	2.28	2.96	2.89	2.89	2.84	5.13	5.05	2.63	2.59

[a] Based on 20 km transportation cost in Table 1. [b] Based on 60 km transportation cost in Table 1. [c] Based on 90% conversion of $CO + H_2$ by the microbial catalyst used and ethanol yield from $CO + H_2$ is 90%. [d] Include supply and labor.

Table 6. Case 2—Comparison of feedstock feed rate, syngas and ethanol yield and ethanol selling prices from gasification and syngas fermentation of 1 kg/s syngas from 5 feedstocks in a 1–2 million gallons per year (MGY) ethanol facility.

Feedstocks Parameters	Wood Chips		Wheat Straw		50 Wheat Straw/50 Manure		OSR Meal		Corn Stover	
	649 °C	850 °C	649 °C	850 °C	649 °C	850 °C	649 °C	850 °C	649 °C	850 °C
Feedstock feed, tonne (db)/d	43.20	37.88	34.95	31.47	29.62	27.23	29.80	27.50	35.93	32.79
Annual demand, thousand tonne (db)/y [a]	12.96	11.36	10.48	9.44	8.87	8.17	8.94	8.25	10.78	9.34
Ethanol production, million gal/y a	1.647	1.513	1.220	1.132	1.039	0.976	1.055	0.992	1.243	1.159
Ethanol yield, gal/tonne (db) [a,b]	127.1	133.1	116.3	119.9	116.9	119.5	118.0	120.2	115.3	117.8
Construction cost, $million	36.35	33.17	31.36	29.13	27.92	26.32	28.04	26.51	31.96	29.98
Operating cost, $million [c]	4.12	3.80	3.62	3.39	3.27	3.11	3.28	3.13	3.68	3.48
Minimum ethanol selling price (MESP), $/gal [a,b,d]	5.63	5.61	7.03	7.04	7.31	7.35	9.49	9.49	6.68	6.74
Minimum ethanol selling price (MESP), $/gal [a,b,e]	5.68	5.65	7.08	7.09	7.35	7.39	9.54	9.54	6.73	6.79

[a] Based on 1 to 2 million gal/year. [b] Based on 90% conversion of $CO + H_2$ by the microbial catalyst used and ethanol yield from $CO + H_2$ is 90%. [c] Include supply and labor. [d] Based on 20 km transportation cost in Table 1. [e] Based on 60 km transportation cost in Table 1.

4. Conclusions

This study showed that integrated commercially available gasification-fermentation systems have the potential to valorize abundant biomass residuals by converting low-valued agricultural and/or forest residuals to liquid fuels. Comparison of syngas production at two output temperatures (649 °C vs. 850 °C) showed that approximately 10% higher feedstock flow rates are required at 649 °C to produce the same syngas flow at 850 °C. Conversion of the four agricultural residuals (i.e., wheat straw, wheat straw/swine manure blend, OSR meal, and corn stover) had up to 11% lower ethanol yields and produced lower amounts of ethanol than that of woody feedstock. The wood chips also had the lowest feedstock costs. Therefore, the lowest MESP was estimated for wood chips at both scales, $2.28 and $5.61 per gallon for the 50 MGY and 1–2 MGY facility, respectively. However, the other three low valued agricultural residuals had MESP similar to wood chips in the 50 MGY facility ($2.59–$2.96), showing that such a facility could be economically operated with a variety of low cost feedstocks. High valued feedstocks such as the OSR meal more than double the MESP at $5.13 per gallon. This demonstrates that the cost for the feedstock plays a considerable role in determining the MESP for the large-scale facility. The biggest difference in MESP was from the two different sized facilities, 50 MGY and 1–2 MGY facilities. The MESP from the smaller (co-op scale, 1–2 MGY) facility were almost twice ($5.61–$9.49/gal) that of larger 50 MGY facility ($2.28–5.13/gal). This highlights the effect of scale-up on construction cost and MESP. With current gasoline selling price of about $2.50 to $3.00/gal, the ethanol production at co-op scale may not be economically sustainable. However, the regional scale ethanol facility may be sustainable by blending with gasoline at comparable prices.

Another key finding of the study is that through a good selection of technologies most farm biomass residuals can be reasonably converted to ethanol via syngas fermentation. Since the largest driving force promoting biofuels, globally and in the USA, are statutory blending requirements [54], the MESP need not be price competitive with gasoline when meeting blending quotas. A more detailed economic evaluation of the market situation is required to determine which range of MESP will be economically sustainable for ethanol use in blended gasolines. Furthermore, the study identified some important challenges: (1) the economic feasibility of smaller scale units should be increased through further research, as well as the development of a pilot scale facility with the gasification-syngas fermentation technology described here. Since the high investment cost of a regional ethanol facility discourages investors, the demonstration of an economically viable smaller scale conversion facility will promote the sustainable use of biomass residuals and offer potential income to farmers. (2) On-going research on alternative products, ranging from higher value niche chemicals to lower cost bulk biofuels from less capital-intensive systems using chemical transformations of syngas (e.g., methanol, dimethyl ether, synthetic natural gas) should continue. Since the feedstock unit costs generally increases with facility size, the optimum facility size for a sustainable product from biomass residuals may very well be a more affordable and smaller facility, most ideally that of the coop size, of 1–2 MGY.

Author Contributions: This research idea was originally conceived by K.S.R. and M.A.D., K.S.R. and W.J.P. analyzed feedstock characteristics of cover crops and manure; M.A.D. and R.P. simulated gasification; H.K.A. and J.A.L. estimated ethanol production; and R.P., H.K.A. and K.S. performed economic analyses. All authors participated in the writing of the manuscript.

Funding: This research received no external funding.

Acknowledgments: The authors would like to acknowledge the technical support by Philip Bauer and Melvin Johnson of the USDA-ARS Coastal Plains Soil, Water & Plant Research Center. This research was supported by the USDA-ARS National Programs 212. Mention of trade names or commercial products in this publication is solely for providing specific information and does not imply recommendation or endorsement by the U.S. Department of Agriculture (USDA).

Conflicts of Interest: The authors declare no conflict of interest.

References

1. Langholtz, M.; Stokes, B.J.; Eaton, L. *2016 Billion-Ton Report: Advanding Domestic Resources for a Thriving Bioeconomy, Volume 1: Economic Availability of Feedstock*; Oak Ridge National Laboratory: Oak Ridge, TN, USA, 2016; p. 448.
2. Dean, J.E.; Weil, R.R. Brassica cover crops for nitrogen retention in the Mid-Atlantic Coastal Plain. *J. Envion. Qual.* **2009**, *38*, 520–528. [CrossRef] [PubMed]
3. SARE. *Managing Cover Crops Profitably: Sustainable Agriculture Research and Eduction (SARE) Handbook Series 9*, 3rd ed.; DIANE Publishing: Collingdale, PA, USA, 2012.
4. Cantrell, K.B.; Ducey, T.F.; Ro, K.S.; Hunt, P.G. Livestock waste-to-bioenergy generation opportunities. *Bioresour. Technol.* **2008**, *99*, 7941–7953. [CrossRef] [PubMed]
5. Ro, K.S.; Libra, J.A.; Bae, S.; Berge, N.D.; Flora, J.R.V.; Pecenka, R. Combustion behavior of animal-manure-based hydrochar and pyrochar. *ACS Sustain. Chem. Eng.* **2019**, *7*, 470–478. [CrossRef]
6. Ro, K.S.; Hunt, P.G.; Jackson, M.A.; Compton, D.L.; Yates, S.R.; Cantrell, K.; Chang, S.C. Co-pyrolysis of swine manure with agricultural plastic waste: Laboratory-scale study. *Waste Manag.* **2014**, *34*, 1520–1528. [CrossRef] [PubMed]
7. Cantrell, K.; Ro, K.S.; Szogi, A.A.; Vanotti, M.B.; Smith, M.C.; Hunt, P.G. Green farming systems for the Southeast USA using manure-to-energy conversion platforms. *J. Renew. Sustain. Energy* **2012**, *4*, 041401. [CrossRef]
8. Hwang, O.; Lee, S.-R.; Cho, S.; Ro, K.S.; Spiechs, M.; Woodbury, B.L.; Silva, P.J.; Han, D.-W.; Choi, H.; Kim, K.-Y.; et al. Efficacy of different biochars in removing odorous volatile organic compounds (VOCs) emitted from swine manure. *ACS Sustain. Chem. Eng.* **2018**, *6*, 14239–14247. [CrossRef]
9. Ro, K.S.; Lima, I.M.; Reddy, G.B.; Jackson, M.A.; Gao, B. Removing gaseous NH3 using biochar as an adsorbent. *Agriculture* **2015**, *5*, 991–1002. [CrossRef]
10. Lima, I.M.; Ro, K.S.; Reddy, G.B.; Boykin, D.L.; Klasson, K.T. Efficacy of chicken litter and wood biochars and their activated counterparts in heavy metal cleanup from wastewater. *Agriculture* **2015**, *5*, 806–825. [CrossRef]
11. Novak, J.M.; Spokas, K.A.; Cantrell, K.; Ro, K.S.; Watts, D.W.; Glaz, B.; Busscher, W.J.; Hunt, P.G. Effects of biochars and hydrochars produced from lignocellulosic and animal manure on fertility of a Mollisol and Entisol. *Soil Use Manag.* **2014**, *30*, 175–181. [CrossRef]
12. Ro, K.S.; Novak, J.M.; Johnson, M.G.; Szogi, A.A.; Libra, J.A.; Spokas, K.A.; Bae, S. Leachate water quality of soils amended with different swine manure-based amendements. *Chemosphere* **2016**, *142*, 92–99. [CrossRef]
13. Cantrell, K.; Ro, K.; Mahajan, D.; Anjom, M.; Hunt, P.G. Role of thermochemical conversion in livestock waste-to-energy treatments: Obstacles and opportunities. *Ind. Eng. Chem. Res.* **2007**, *46*, 8918–8927. [CrossRef]
14. INEOS. Available online: https://www.ineos.com/news/ineos-group/ineos-bio-produces-cellulosic-ethanol-at-commercial-scale/ (accessed on 31 July 2013).
15. LanzaTech. Available online: https://www.lanzatech.com/ (accessed on 2 July 2019).
16. Kirsanovs, V.; Zandeckis, A.; Blumberga, D.; Veidenbergs, I. Influence of process temperature, equivalence ratio and fuel moisture content on gasification process, a review. In Proceedings of the 27th International Conference on Efficiency, Cost, Optimization, Simulation and Environmental Impact of Energy Systems—ECOS, Turku, Finland, June 2014.
17. Ince, P.; Bilek, E.; Dietenberger, M. Modeling integrated biomass gasification business concepts. **2011**. [CrossRef]
18. Dietenberger, M.; Anderson, M. Vision of the U.S. biofuel future: A case for hydrogen-enriched biomass gasification. *Ind. Eng. Chem. Res.* **2007**, *46*, 8863–8874. [CrossRef]
19. Wu, Y.; Yang, W.; Blasiak, W. Energy and exergy analysis of high temperature agent gasification of biomass. *Energies* **2014**, *7*, 2107–2122. [CrossRef]
20. Pian, C.C.P.; Yoshikawa, K. Development of a high-temperature air-blown gasification system. *Bioresour. Technol.* **2001**, *79*, 231–241. [CrossRef]
21. Young, L.; Pian, C.C.P. High-temperature, air-blown gasification of dairy-farm wastes for energy product. *Energy* **2003**, *28*, 655–672. [CrossRef]

22. Kopke, M.; Mihalcea, C.; Bromley, J.C.; Simpson, S.D. Fermentative production of ethanol from carbon monooxide. *Curr. Opin. Biotechnol.* **2011**, *22*, 320–325. [CrossRef]
23. Phillips, J.; Huhnke, R.; Atiyeh, H. Syngas Fermentation: A Microbial Conversion Process of Gaseous Substrates to Various Products. *Fermentation* **2017**, *3*, 28. [CrossRef]
24. Abubackar, H.N.; Veiga, M.C.; Kennes, C. Carbon monoxide fermentation to ethanol by Clostridium autoethanogenum in a bioreactor with no accumulation of acetic acid. *Bioresour. Technol.* **2015**, *186*, 122–127. [CrossRef]
25. Devarapalli, M.; Atiyeh, H.K.; Phillips, J.R.; Lewis, R.S.; Huhnke, R.L. Ethanol production during semi-continuous syngas fermentation in a trickle bed reactor using Clostridium ragsdalei. *Bioresour. Technol.* **2016**, *209*, 56–65. [CrossRef]
26. Huhnke, R.L.; Lewis, R.S.; Tanner, R.S. Isolation and characterization of novel clostridial species. U.S. Patent 7,704,723, 27 April 2010.
27. Liu, K.; Atiyeh, H.K.; Tanner, R.S.; Wilkins, M.R.; Huhnke, R.L. Fermentative production of ethanol from syngas using novel moderately alkaliphilic strains of Alkalibaculum bacchi. *Bioresour. Technol.* **2012**, *104*, 336–341. [CrossRef]
28. Phillips, J.R.; Atiyeh, H.K.; Tanner, R.S.; Torres, J.R.; Saxena, J.; Wilkins, M.R.; Huhnke, R.L. Butanol and hexanol production in Clostridium carboxidivorans syngas fermentation: Medium development and culture techniques. *Bioresour. Technol.* **2015**, *190*, 114–121. [CrossRef]
29. Phillips, J.R.; Klasson, K.T.; Clausen, E.C.; Gaddy, J.L. Biological production of ethanol from coal synthesis gas. *Appl. Biochem. Biotechnol.* **1993**, *39*, 559–571. [CrossRef]
30. Sun, X.; Atiyeh, H.K.; Zhang, H.; Tanner, R.S.; Huhnke, R.L. Enhanced ethanol production from syngas by Clostridium ragsdalei in continuous stirred tank reactor using medium with poultry litter biochar. *Appl. Energy* **2019**, *236*, 1269–1279. [CrossRef]
31. Diender, M.; Stams, A.J.M.; Sousa, D.Z. Production of medium-chain fatty acids and higher alcohols by a synthetic co-culture grown on carbon monoxide or syngas. *Biotechnol. Biofuels* **2016**, *9*, 82. [CrossRef]
32. He, P.; Han, W.; Shao, L.; Lü, F. One-step production of C6–C8 carboxylates by mixed culture solely grown on CO. *Biotechnol. Biofuels* **2018**, *11*, 4. [CrossRef] [PubMed]
33. Liu, K.; Atiyeh, H.K.; Stevenson, B.S.; Tanner, R.S.; Wilkins, M.R.; Huhnke, R.L. Mixed culture syngas fermentation and conversion of carboxylic acids into alcohols. *Bioresour. Technol.* **2014**, *152*, 337–346. [CrossRef]
34. Richter, H.; Molitor, B.; Diender, M.; Sousa, D.Z.; Angenent, L.T. A Narrow pH Range Supports Butanol, Hexanol, and Octanol Production from Syngas in a Continuous Co-culture of Clostridium ljungdahlii and Clostridium kluyveri with In-Line Product Extraction. *Front. Microbiol.* **2016**, *7*, 1773. [CrossRef]
35. Wang, Y.-Q.; Zhang, F.; Zhang, W.; Dai, K.; Wang, H.-J.; Li, X.; Zeng, R.J. Hydrogen and carbon dioxide mixed culture fermentation in a hollow-fiber membrane biofilm reactor at 25 °C. *Bioresour. Technol.* **2018**, *249*, 659–665. [CrossRef]
36. Xu, S.; Fu, B.; Zhang, L.; Liu, H. Bioconversion of H_2/CO_2 by acetogen enriched cultures for acetate and ethanol production: The impact of pH. *World J. Microbiol. Biotechnol.* **2015**, *31*, 941–950. [CrossRef]
37. Werpy, T.; Petersen, G. *Top Value Added Chemicals from Biomass: Volume 1—Results of Screening for Potential Candidates from Sugars and Synthesis Gas*, 1st ed.; National Renewable Energy Laboratory: Golden, CO, USA, 2004.
38. Pfltzgraff, L.A.; De bruyn, M.; Cooper, E.C.; Budarin, V.; Clark, J.H. Food waste biomass: A resource for high-value chemicals. *Green Chem.* **2013**, *15*, 307–314. [CrossRef]
39. Molino, A.; Larocca, V.; Chianese, S.; Musmarra, D. Biofuels production by biomass gasification: A review. *Energies* **2018**, *11*, 811. [CrossRef]
40. Bergman, R.; Berry, M.; Bilek, E.M.T.; Bower, T.; Eastin, I.; Ganguly, I.; Han, H.-S.; Hirth, K.; Jacobson, A.; Karp, S.; et al. Waste to Wisdom: Utilizing forest residues for the production of bioenergy and biobased products. *Appl. Eng. Agric.* **2018**, *34*, 5–10.
41. Sahoo, K.; Bilek, E.; Bergman, R.; Mani, S. Techno-economic analysis of producing solid biofuels and biochar from forest residues using portable systems. *Appl. Energy* **2019**, *235*, 578–590. [CrossRef]
42. Sahoo, K.; Mani, S. Engineering Economics of Cotton Stalk Supply Logistics Systems for Bioenergy Applications. *Trans. ASABE* **2016**, *59*, 737–747.

43. Sahoo, K.; Mani, S. Techno-economic assessment of biomass bales storage systems for a large-scale biorefinery. *Biofuels Bioprod. Biorefin.* **2017**, *11*, 417–429. [CrossRef]
44. Sahoo, K.; Bilek, E.M.; Mani, S. Techno-economic and environmental assessments of storing woodchips and pellets for bioenergy applications. *Renew. Sustain. Energy Rev.* **2018**, *98*, 27–39. [CrossRef]
45. Sahoo, K.; Bilek, E.M.; Bergman, R.D.; Kizha, A.R.; Mani, S. Economic analysis of forest residues supply chain options to produce enhanced quality feedstocks. *Biofuels Bioprod. Biorefin.* **2018**, *13*, 514–534. [CrossRef]
46. USDA Oil crops yearbook. Available online: https://www.ers.usda.gov/data-products/oil-crops-yearbook/ (accessed on 15 June 2019).
47. Sahoo, K. Sustainable Design and Simulation of Multi-Feedstock Bioenergy Supply Chain. Doctroal Thesis, University of Georgia, Athens, Greece, 2017; 436p.
48. Eriksson, G.; Hedman, H.; Bostrom, D.; Pettersson, E.; Backman, R.; Ohman, M. Combustion characterization of rapeseed meal and possible combustion applications. *Energy Fuels* **2009**, *23*, 3930–3939. [CrossRef]
49. Ro, K.S.; Cantrell, K.B.; Hunt, P.G. High-temperature pyrolysis of blended animal manures for producing renewable energy and value-added biochar. *Ind. Eng. Chem. Res.* **2010**, *49*, 10125–10131. [CrossRef]
50. Ljungdhal, L. The autotrophic pathway of acetate synthesis in acetogenic bacteria. *Annu. Rev. Microbiol.* **1986**, *40*, 415–450. [CrossRef]
51. Köpke, M.; Mihalcea, C.; Liew, F.; Tizard, J.H.; Ali, M.S.; Conolly, J.J.; Al-Sinawi, B.; Simpson, S.D. 2,3-Butanediol Production by Acetogenic Bacteria, an Alternative Route to Chemical Synthesis, Using Industrial Waste Gas. *Appl. Environ. Microbiol.* **2011**, *77*, 5467–5475. [CrossRef]
52. Fernández-Naveira, Á.; Veiga, M.C.; Kennes, C. H-B-E (hexanol-butanol-ethanol) fermentation for the production of higher alcohols from syngas/waste gas. *J. Chem. Technol. Biotechnol.* **2017**, *92*, 712–731. [CrossRef]
53. Xu, D.; Tree, D.R.; Lewis, R.S. The effects of syngas impurities on syngas fermentation to liquid fuels. *Biomass Bioenergy* **2011**, *35*, 2690–2696. [CrossRef]
54. UFOP. UFOP Report on Global Market Supply 2017/2018. Available online: https://www.ufop.de/files/3515/1515/2657/UFOP_Report_on_Global_Market_Supply_2017-2018.pdf (accessed on 16 August 2019).

environments

MDPI

Article

Application of Gas-Permeable Membranes For-Semi-Continuous Ammonia Recovery from Swine Manure

Berta Riaño [1,*], Beatriz Molinuevo-Salces [1], Matías B. Vanotti [2] and María Cruz García-González [1]

[1] Agricultural Technological Institute of Castilla y Léon. Ctra. Burgos, km. 119, 47071 Valladolid, Spain; ita-molsalbe@itacyl.es (B.M.-S.); gargonmi@itacyl.es (M.C.G.-G.)

[2] United States Department of Agriculture, Agricultural Research Service, Coastal Plains Soil, Water and Plant Research Center, 2611, W. Lucas St., Florence, SC 29501, USA; Matias.Vanotti@ars.usda.gov

* Correspondence: riairabe@itacyl.es; Tel.: +34-983-317-384

Received: 30 January 2019; Accepted: 3 March 2019; Published: 6 March 2019

Abstract: Gas-permeable membrane technology is a new strategy to minimize ammonia losses from manure, reducing pollution and recovering N in the form of an ammonium salt fertilizer. In this work, a new operational configuration to recover N using the gas-permeable membrane technology from swine manure was tested in a semi-continuous mode. It treated swine manure with a total ammonia nitrogen (TAN) concentration of 3451 mg L^{-1}. The system was operated with low aeration rate (to raise pH), and with hydraulic retention times (HRT) of seven days (Period I) and five days (Period II) that provided total ammonia nitrogen loading rate (ALR) treatments of 491 and 696 mg TAN per L of reactor per day, respectively. Results showed a uniform TAN recovery rate of 27 g per m^2 of membrane surface per day regardless of the ALR applied and the manure TAN concentration in the reactor. TAN removal reached 79% for Period I and 56% for Period II, with 90% of recovery by the membrane in both periods. Water capture in the acidic solution was also uniform during the experimental period. An increase in temperature of 3 °C of the acidic solution relative to the wastewater reduced 34% the osmotic distillation and water dilution of the product. These results suggested that the gas-permeable membrane technology operating in a semi-continuous mode has a great potential for TAN recovery from manure.

Keywords: total ammonia nitrogen; recovery; swine manure; gas-permeable membranes

---

1. Introduction

Ammonia (NH_3) is a cause of air pollution and can potentially contribute to acidification and eutrophication, both of which can damage sensitive vegetation, biodiversity, water quality and human health [1–3]. The agricultural sector was the responsible for 93% of NH_3 emissions in the European Union (EU) in 2013, resulting in 3.6 million tons. NH_3 volatilization from livestock wastes accounted for almost 64% of the agricultural NH_3 emissions [2]. Significant efforts are required to abate NH_3 emissions from agricultural sources, mainly those coming from livestock wastes [4]. On the other hand, current practices used for production of nitrogen (N) fertilizers via the Haber–Bosch process are cost and energy intensive and contribute to global warming [5,6]. There is a renewed interest in recent years to recover nutrients from waste streams due to a combination of economic, environmental, and energy considerations [7,8].

Different technologies have been investigated for capture and recovery of NH_3 emissions from livestock wastes. These technologies include: reverse osmosis using high pressure and hydrophilic membranes [9], ammonia stripping using stripping towers and acid absorption [10], zeolite adsorption through ion exchange [11], struvite precipitation through co-precipitation with phosphate and magnesium [12], and, more recently, gas-permeable membrane technology [13]. Traditional processes present some limitations: (1) reverse osmosis requires high pressure; (2) air stripping towers and zeolite adsorption techniques need pre-treatment of manure; and (3) struvite precipitation requires the addition of $Mg^{2+}$and PO_4^{3+} to balance the stoichiometry of struvite precipitation [14]. The technology based on gas-permeable membranes presents several advantages over traditional processes such as (1) low energy consumption (0.18 kWh kg NH_3^{-1}), (2) it is carried out at low pressure, (3) it does not require pre-treatment of wastewater, and (4) it does not need addition of any alkali reagent [15–17].

The most important phenomenon related to gas-permeable membranes is the mass transfer driven by the difference in NH_3 gas concentration between both sides of the microporous, hydrophobic membrane [14,18]. More specifically, NH_3 contained in the livestock wastes passes through the membrane, being captured and concentrated in an acidic stripping solution on the other side of the membrane. The efficiency of the gas-permeable membrane is directly related to the availability of NH_3 in the waste, where the total ammonia nitrogen species (TAN) NH_3 and NH_4^+, are in equilibrium [19–21]. This equilibrium depends on the pH and temperature of the livestock waste, having the pH a greater influence [15,21]. Alkaline pH causes dissociation of NH_4^+ and forms free NH_3 that can cross the membrane and be captured by the acidic solution.

The gas-permeable membrane technology has been successfully applied to recover up to 99% of TAN from swine manure and anaerobically digested swine manure [5,14,21–23]. Previous research mainly focused on the influence of operational conditions such as animal waste strength and pH on TAN recovery using gas-permeable membranes always operated at batch mode. However, there is no experience operating this type of system in a semi-continuous mode. Thus, gathering more experience and experimental data on the use of gas-permeable membranes with new configurations for recovering TAN from livestock wastes is of major importance towards the development and demonstration of this technology.

The objective of this study was to determine TAN recovery from swine manure using a gas-permeable membrane system operating at semi-continuous mode. The semi-continuous gas-permeable system was tested with decreasing hydraulic retention times (HRT) from seven to five days and increasing total ammonia nitrogen loading rates (ALR) in the range of 38.5 and 54.6 g TAN per m^2 of membrane surface per day. The system was monitored in terms of removal and recovery of TAN as well as in changes of organic matter and solids in the swine manure during operation.

2. Materials and Methods

2.1. Origin of Manure

Swine manure was collected from a farm located in Narros de Cuellar (Segovia, Spain). The manure was a centrate collected after on-farm centrifugation. The mean concentrations for centrifuged manure were pH of 7.6 ± 0.2, 33.3 ± 3.5 g total solids (TS) L^{-1}, 23.5 ± 3.2 g volatile solids (VS) L^{-1}, 67.1 ± 10.1 g total chemical oxygen demand (CODt) L^{-1}, 3451 ± 132 mg TAN L^{-1}, 253 ± 59 mg total phosphorous (TP) L^{-1} and 3119 ± 4 mg potassium L^{-1}. The liquid centrifuged manure was collected in plastic containers, transported in coolers to the laboratory and subsequently stored at 4 °C for further use.

2.2. Semi-Continuous Recovery of Ammonia from Manure

2.2.1. Experimental Set-Up

The experimental set-up consisted of a reactor with a total working volume of 2 L of fresh centrifuged swine manure (30 cm long, 20 cm wide, 4 cm high) (Figure 1). The acid tank used to concentrate TAN consisted of a 500 mL Erlenmeyer flask containing an acidic solution (300 mL of 1 N H_2SO_4) which was replaced by a bigger flask. This was due to the increase of the volume of the acidic solution with time as a consequence of the occurrence of osmotic distillation (OD).

This acidic solution was continuously recirculated using a peristaltic pump (Pumpdrive 5001, Heidolph, Schwabach, Germany) at 11 L d^{-1} through a tubular gas-permeable membrane submerged in the reactor. The tubular membrane was made of expanded polytetrafluoroethylene (e-PTFE) material (Zeus Industrial Products Inc., Orangeburg, SC, USA). Membrane specifications are provided in Table 1. The ratio of the e-PTFE membrane length per effective reactor volume was 0.4 m L^{-1} and the ratio of the e-PTFE membrane area per reactor volume was 0.013 m^2 L^{-1}. The tubular membrane was placed in a bended horizontal configuration and held submerged by plastic connections. Low-rate aeration was used to naturally increased manure pH without chemicals according to previous work [15]. Air was supplied using an aquarium air pump (Hailea, Aco-2201) from the bottom of the reactor through a porous stone. The airflow rate was controlled at 0.24 L-air L $manure^{-1}$ min^{-1} using an airflow meter (Aalborg, Orangeburg, NY, USA). The lid of the reactors was not sealed, having one open port that allowed air to escape. In order to ensure nitrification inhibition, a nitrification inhibitor (allythiourea) was added to the manure at a concentration of 10 mg L^{-1}. The manure was continuously agitated using magnetic stirrers.

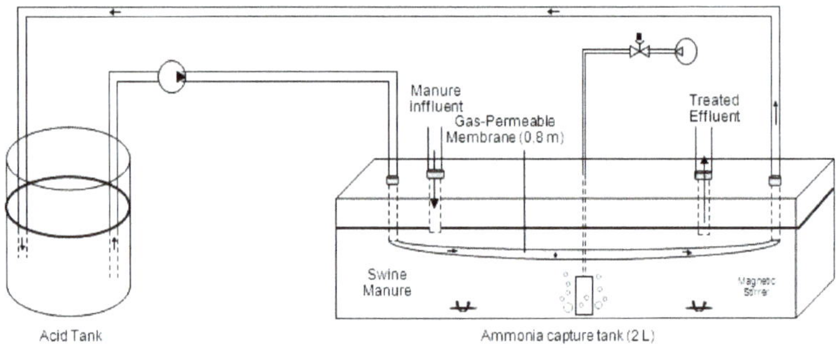

Figure 1. Process diagram of gas-permeable system operating at semi-continuous mode.

Table 1. Specifications of the gas-permeable membranes used in experiments described in Sections 2.2 and 2.3.

Membranes Properties	First Experiment (Section 2.2)	Second Experiment (Section 2.3)
Inner diameter (mm)	8.64	4.08
Wall thickness (mm)	0.76	0.56
Pore size (μm)	2.5	-
Bubble point (kPa)	207	-
Density (g cm^3)	0.45	0.95
Length (m)	0.8	0.61
Surface area (m^2)	0.026	0.091

The volume of the reactor was checked daily and any water lost was replenished. The reactor was manually fed in semi-continuous mode five times per week: from Monday to Thursday with a load equivalent to one day each day, and Fridays with a load equivalent to three days. Prior to each feeding event, a volume equal to the feeding volume was removed from the reactor. The reactor was fed at an hydraulic retention time (HRT) of 7 d (total 2.0 L per week) during Period I (1–30 d) and at a HRT of 5 d (total 2.8 L per week) during Period II (31–50 d), corresponding to total ammonia nitrogen loading rates (ALR) of 491 and 696 mg TAN per L of the reactor per day (i.e., 38.5 and 54.6 g TAN per m^2 of membrane surface per day, respectively). As TAN was depleted from manure and transferred to the acid tank, the pH of the acidic solution increased. A protocol was established: concentrated H_2SO_4 (96–98%, Panreac) was added to the acidic solution to an endpoint of pH < 1 whenever the pH of the acidic solution increased up to 2. The experiment was performed at a temperature of 22.0 ± 1.7 °C in duplicate reactors and the results were expressed as means and standard deviations.

2.2.2. Manure and Acidic Solution Sampling

Samples of 50 mL were daily taken from the influent and from the effluent of the reactor. pH and TAN concentration were determined daily in both the influent and the effluent of the reactor. For each experimental day, TAN removal efficiency was calculated using Equation (1):

$$\text{TAN removal efficiency (\%)} = 100 \times (\text{TAN}_{\text{in}} - \text{TAN}_{\text{eff}})/\text{TAN}_{\text{eff}} \quad (1)$$

where TAN_{in} is influent TAN concentration and TAN_{eff} is TAN concentration in the effluent for each experimental day.

Total alkalinity in influent and effluent was also determined twice a week, and analyses of total Kjeldahl nitrogen (TKN), TS, VS, and CODt were performed on samples collected once a week. Acidic solution samples of 6 mL from the acid tank were also collected daily to monitor pH and TAN. The acidic solution was analyzed at the end of the experiment for conductivity, CODt, total volatile fatty acids (TVFA), TP, sulfur, potassium, magnesium, calcium, zinc, copper, and iron.

2.3. *Effect of Differential Heating on Osmotic Distillation*

Two assays were performed in order to evaluate the effect of differential heating of the acidic solution on osmotic distillation, which is the passage of water vapor through the gas-permeable membrane from the wastewater into the acid trap. In the first assay (control assay) the acidic solution was not heated. The measured temperature of the acidic solution was 23.2 ± 0.5 °C and that of the wastewater in the reactor was 24.3 ± 0.8 °C. In the second assay, the temperature of the acidic solution (30.0 ± 0.8 °C) was kept at 3 °C above the temperature of the reactor (26.9 ± 0.8 °C) using a heated water bath.

For these assays, the experimental set-up consisted of a reactor with a total working volume of 0.7 L (diameter 20 cm, height 3 cm). In this case, a synthetic solution having a balanced TAN to alkalinity ratio > 4.1 [24] was used to simulate swine manure, consisting of NH_4Cl at a concentration of 3.47 ± 0.07 g TAN L^{-1} and $NaHCO_3$ at a concentration of 15.6 ± 0.2 g $CaCO_3$ L^{-1}. In order to ensure nitrification inhibition, a nitrification inhibitor (allythiourea) was added at a concentration of 10 mg L^{-1}. Average pH of this synthetic solution was 8.3 ± 0.3. The tank used to concentrate TAN in each reactor consisted of a 500 mL Erlenmeyer flask initially containing 90 mL of 1 N H_2SO_4. This stripping solution was continuously recirculated using a peristaltic pump (Pumpdrive 5001, Heidolph, Schwabach, Germany) through a tubular gas-permeable membrane submerged in the reactors at a flow rate of 12 L d^{-1}. The tubular membrane was made of e-PTFE material (Zeus Industrial Products Inc., Orangeburg, SC, USA) and their characteristics are provided in Table 1. The ratio of the tubular membrane length per synthetic wastewater volume was 0.8 m L^{-1} and the ratio of the membrane area per volume of synthetic wastewater was 0.013 m^2 L^{-1}. Reactors were fed at a HRT of 7 d during a period of 7 d each assay. The evaluation of the performance of the system and the sampling method was identical to those described in Sections 2.2.1 and 2.2.2, respectively. Each assay was conducted in duplicate.

2.4. *Analytical Methods and Statistical Analysis*

Total alkalinity and pH were monitored using a pH meter Crison Basic 20 (Crison Instruments S.A., Barcelona, Spain). Total alkalinity was determined by measuring the amount of standard sulfuric acid needed to bring the sample to pH of 4.5. Analyses of TS, VS, CODt, TAN, and TP were performed in accordance with Standard Methods [25], according to methods 2540 B for VS, 5220-D for CODt, 4500-NH_3 E for TAN and 4500-P C for TP.

Conductivity was measured using a conductimeter Crison 524 (Crison Instruments S.A., Barcelona, Spain). Magnesium, calcium, zinc, copper, and iron were analyzed using an atomic absorption spectrometer (AA 240 FS, Varian). Potassium was analyzed using an atomic emission spectrometer (AA 240 FS, Varian). These compounds were analysed following the methods described by USEPA [26]:

for magnesium EPA method 215.1, for calcium EPA method 242.1; for zinc EPA method 289.1, for iron EPA method 236.1 and for potassium EPA 258.1. Sulfur was measured by combustion and infrared detection (LECO CNS 2000). The concentration of TVFA (i.e., sum of acetic, propionic, butyric, iso-butyric, valeric, iso-valeric, hexanoic and heptanoic acids) was determined using a gas chromatograph (Agilent 7890A) equipped with a Teknokroma TRB-FFAP column of 30 m length and 0.25 mm i.d. followed by a flame ionization detector (FID). The carrier gas was helium (1 mL min^{-1}). The temperature of the detector and the injector was 280 °C. The temperature of the oven was set at 100 °C for 4 min, then increased to 150 °C for 2 min and thereafter increased to 210 °C.

Free ammonia (FA) was quantified according to Hansen et al. [27] (Equation (2)):

$$[NH_3]/[TAN] = (1 + (10^{-pH}/10^{-(0.09018 + 2729.92/T)}))^{-1} \quad (2)$$

where NH_3 was the FA content, T was the manure reactor temperature, and pH was measured in the effluent.

The mass transfer coefficient (K_m; m d^{-1}) has been calculated using Equation (3) [18]:

$$J = K_m (C_1 - C_2) \quad (3)$$

where J is the TAN mass flux per area (g m^{-2} d^{-1}), and C_1 and C_2 are the concentrations of free ammonia. The K_m coefficient depends on several factors, including the flow rate of the acidic solution and membrane characteristics such as porosity or thickness [18].

Results obtained were analysed using one-way analysis of variance (ANOVA) with significance at $p < 0.05$.

3. Results and Discussion

3.1. TAN Removal and Recovery by the Gas-Permeable System in Semi-Continuous Mode: Effect of Total Ammonia Nitrogen Loading Rate

In this study, the TAN removal from centrifuged swine manure was evaluated using a gas-permeable membrane with low-rate aeration at semi-continuous mode. The semi-continuous system was evaluated with two different ALRs: 491 mg TAN L^{-1} d^{-1} for Period I (1–30 d) and 696 mg TAN L^{-1} d^{-1} for Period II (31–50 d). Corresponding HRTs were 7 d during Period I and 5 d during Period II. As shown in Figure 2, TAN concentration in the manure effluent decreased steadily from 3646 mg L^{-1} to 611 mg L^{-1} in the first 16 days. After this day, TAN concentration remained approximately constant (average value of 690 ± 139 mg L^{-1}). When ALR was increased in Period II, TAN concentration in the effluent significantly ($p < 0.05$) increased up to a constant concentration of about 1500 mg L^{-1}. Free ammonia concentration during the process inside the reactor (calculated according to Equation (2)) also varied between periods with average values of 195 ± 61 mg L^{-1} for Period I and 328 ± 56 mg L^{-1} for Period II. As an average, TAN removal was 79 ± 5% for Period I (days 10–30) and 56 ± 7% for Period II (days 35–50). However, the rate of TAN mass removal (mg TAN per day) from manure was the same during the whole experimentation time regardless of the ALR applied, as it can be seen from the uniform linear trend ($R^2 = 0.9962$) of the mass TAN removal vs. time data presented in Figure 3 that fits well both periods. No significant differences ($p = 0.41$) were found among TAN removal rates of the two periods. The slope present in Figure 3 leads to a TAN removal rate of 387 mg L^{-1} d^{-1}.

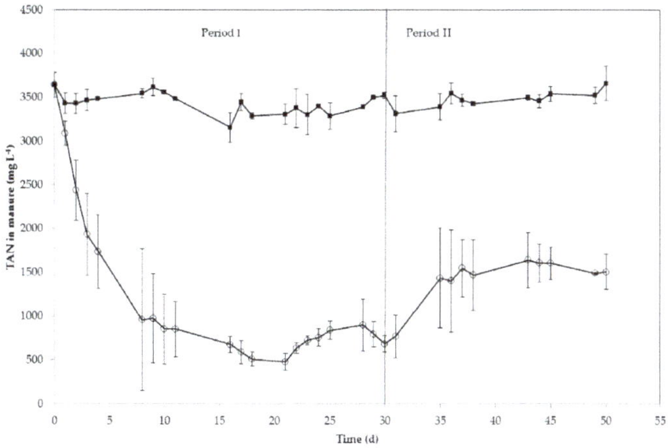

Figure 2. TAN concentration in the influent swine manure (■) and in effluent of the gas-permeable membrane system (○). The error bars represent the standard deviation of duplicate experiments.

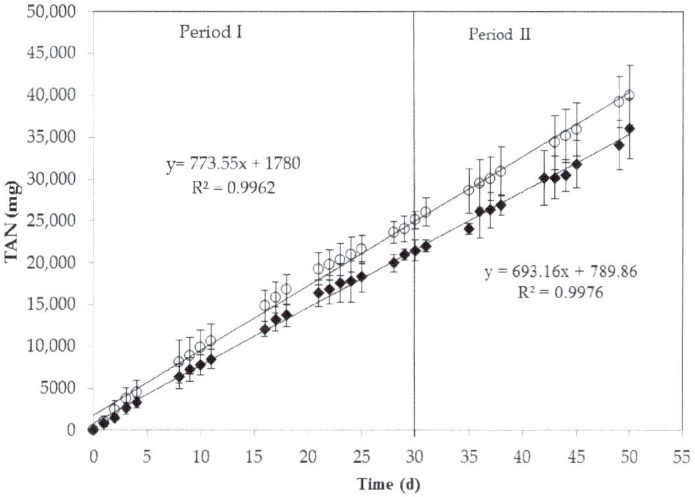

Figure 3. Mass of TAN removed from swine manure (○) and recovered in the acid tank (♦). Linear equations and R^2 are represented. The error bars represent the standard deviation of duplicate experiments.

The results displayed the same trend for TAN recovery in the acidic solution (Figure 3). A linear (R^2 = 0.9976) TAN recovery rate by the gas-permeable membrane with a mass recovery rate of 347 mg L^{-1} d^{-1} (27.1 g m^{-2} d^{-1}), and an average TAN recovery efficiency of 90%, were evidenced during the whole experimentation time regardless of ALR and TAN content in reactor (Figure 3). No significant differences (p = 0.71) in TAN recovery rates was observed between both periods. The mass transfer coefficient (k_m) was calculated according to Equation (3), resulting in an average value of 1.5×10^{-6} m s^{-1}. This value was in the range of that reported by Samani-Majd et al. [18] for an e-PTFE membrane system operating at different pH values. The loss of TAN by the system (mass difference between TAN removal and TAN recovery) was 10% of TAN removal and could be due to volatilized N loss as the system was not closed. A nitrogen mass balance was carried out and no organic nitrogen degradation was evidenced (Table 2), thus an increase of initial TAN concentration did not occur due to biological activity.

Table 2. Nitrogen mass balance for the system for Period I and Period II.

Period	TKN_{in} (mg L^{-1})	TAN_{in} (mg L^{-1})	N org_{in} (mg L^{-1})	TKN_{eff} (mg L^{-1})	TAN_{eff} (mg L^{-1})	N org_{eff} (mg L^{-1})
I	4481 ± 182	3425 ± 122	1029	1830 ± 381	748 ± 153	1082
II	4685 ± 222	3486 ± 134	1199	2745 ± 345	1525 ± 83	1220

TKN_{in} is influent TKN concentration; $Norg_{in}$ is the influent organic N concentration; TKN_{eff} is the TKN concentration in the effluent; $Norg_{eff}$ is the organic N concentration in the effluent.

Although the TAN concentration in the reactors varied significantly between periods, the TAN mass recovery rate was constant. This is surprising because previous research in batch systems indicated a marked effect of TAN concentration on the mass N recovery rate by the gas-permeable membrane system [14]. Table 3 shows a comparison of recovery efficiencies obtained by other authors operating gas-permeable membranes systems in batch mode and the results of this study operating in semi-continuous mode. Operating at batch mode, Vanotti et al. [28] treated anaerobically digested swine wastewater containing 2350 mg TAN L^{-1} using submerged membranes plus low-rate aeration to recover NH_3. TAN reduction obtained at five to six days was higher than 93%. Dube et al. [5] studied the effect of aeration on pH increased when treating digested effluents from covered anaerobic swine lagoons with different TAN concentrations. The pH of digested effluents with aeration increased to 9.2, achieving TAN recovery efficiencies of 96–98% in five days of batch operation. In that study, the recovery of TAN was five times faster with aeration compared with treatment without aeration. García-González et al. [15] also tested in batch mode operation the application of low-rate aeration as an alternative to the use of alkali to recover NH_3 from raw swine manure with 2390 mg TAN L^{-1}. Under these conditions, the manure pH increased above 8.5 with 99% TAN recovery efficiency. Table 3 shows the comparison between these studies and the present study all based on the average TAN recovery rate per membrane area (in g of TAN m^{-2} membrane d^{-1}). The average TAN recovery in the semi-continuous system accounted for 27.1 g m^{-2} d^{-1} that was in the range (22.7–30.7 g m^{-2} d^{-1}) of recoveries obtained in batch studies with similar treatment time (Table 3). This good performance in semi-continuous mode was obtained with a process pH of 8.46 that was about one unit lower than the pH obtained with aeration using batch mode (up to 9.5) (Table 3). pH was the most critical variable determining the amount of free NH_3 available to pass through the gas-permeable membrane [14]. In batch mode, ammonia capture efficiency decreased as TAN concentration was depleted from the reactors [14]. However, in semi-continuous mode the daily supply of TAN maintained a consistently high TAN recovery rate in spite of the lower pH process.

Table 3. Comparison of results from this study operating in semi-continuous mode with previous studies also recovering TAN using gas-permeable membranes from livestock wastewaters applying low-rate aeration but operating in batch mode.

Type of Wastewater	Operation Mode [a]	Treatment Time (d)	Ratio Membrane Surface/Manure Volume (m^2 L^{-1})	Aeration Rate (L_{air} L_{waste}^{-1} min^{-1})	Initial TAN Concentration (mg L^{-1})	TAN Removal (%)	TAN Recovery over Removed (%)	Average TAN Recovery (g m^{-2} d^{-1}) [b]	Initial pH	Final pH	Reference
Digested swine effluent	Batch	6	0.013	0.12	2350	97	93 [c]	25.1 [c]	8.36	9.47	[28]
Anaerobically digested swine manure	Batch	5	0.013	0.12	2097	97	98	30.7	8.71	9.26	[5]
Anaerobically digested swine manure	Batch	5	0.013	0.12	1465	99	96	22.7	8.47	9.17	[5]
Raw swine manure	Batch	18	0.013	0.24	2390	99	99	9.5	7.50	9.20	[15]
Raw swine manure centrate	Semi-continuous	7	0.013	0.24	3451	79	90	27.1	7.60	8.46	Present study (period I)

[a] Operational temperatures were laboratory room temperatures between 22–25 °C in all studies. [b] TAN recovery efficiency was equal to TAN recovered in the acidic solution divided by TAN removed from manure multiplied by 100. [c] 7.8% TAN was recovered in phosphorous precipitate.

3.2. Characterization of the Acidic Solution Containing the Concentrated Ammonia Product

TAN concentration in the acidic solution rapidly increased in the first 16 days to values near 16,000 mg L^{-1}, increasing much slower after that time and reaching an approximately constant concentration from day 21 close to 19,000 mg L^{-1} (Figure 4). This is attributed to the diffusion of water vapor gas through the membrane, averaging 16 g per day and liter of manure in reactor, corresponding to 1252 g of water per day and m^2 of membrane surface (Figure 4). Similar to TAN recovery, water capture in the acidic solution was uniform during the whole experimental period, with no significant differences between the two periods ($p = 0.53$). This led to a continuous dilution of the acidic solution and increased its volume. Thus, the weight of the acidic solution was more than 6-fold higher after 50 days of experimentation compared to the initial weight (Figure 4). Darestani et al. [29] pointed out that the transfer of water vapor may occur during the TAN removal process using hydrophobic membranes due to differences in vapor pressure between both sides of the membrane (i.e., osmotic distillation or OD). This process could have a great impact on the economy of the process. Firstly, TAN could not be concentrated in the acidic solution as expected and a further process will be required to concentrate ammonium sulfate and to reduce transportation cost to export this fertilizer outside the farm. Secondly, OD also affects the design of the acid tank that must foresee increasing volume of the acidic solution. A possible strategy to counteract and effectively inhibit osmotic distillation could be heating the stripping solution and/or cooling the feed solution [19,30]. Therefore, this strategy was evaluated with a new assay shown in Table 4. A temperature increment of the acidic solution of only 3 °C caused a decrease of approximately 34% of the OD as represented by the reduced water recovery in the acidic solution, with no significant differences in ammonia recovery (Table 3). Thus, heating of the acidic tank by a few °C offers a good and cheap alternative for reducing OD during recovery of ammonia using gas-permeable membranes.

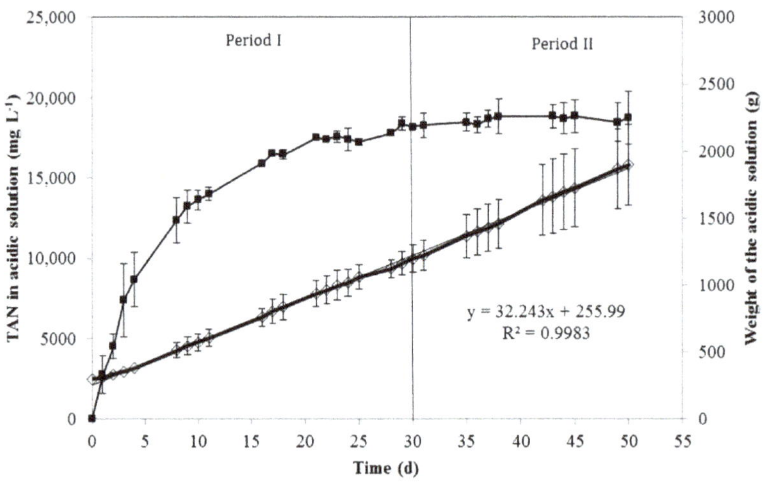

Figure 4. TAN concentration in the acid tank (■) and weight of the acidic solution (◇) during the experimental time. The error bars represent the standard deviation of duplicate experiments.

Table 4. Effect of heating the acidic solution (3 °C warmer than wastewater in the reactor) on the recovery of both water and ammonia. Data are means of duplicate experiments ± standard deviations.

	No heating	Heating
Water recovery (g m^{-2} d^{-1})	921 ± 185	612 ± 22
Ammonia recovery (g m^{-2} d^{-1})	49.5 ± 0.3	51.8 ± 2.6

The acidic solution contained about 24% (dry weight) of sulfur, which is also a valuable nutrient for fertilization. Regarding the presence of other inorganic compounds, among potassium, magnesium, calcium, zinc, copper, and iron, only K^+ was detected in average value of 28 mg L^{-1}. The high salt concentrations lead to a conductivity of the acidic solution of 90 mS cm^{-1}.

After some days of testing, the acidic solution changed from transparent to brownish. The same visual observation was made by Zarebska et al. [31]. The quantification of COD concentration of this solution indicated a low level (139 $\pm$ 12 mg L^{-1}). Such a result is consistent with the analysis of total VFA by gas chromatography, in which only acetic acid was detected among seven VFA determinations, at a concentration of 60 $\pm$ 25 mg acetic acid L^{-1} equivalent to 64 mg COD L^{-1} (using a conversion factor of 1.07 [32]). Xie et al. [8] indicated that volatile organic compounds, such as VFA that exert partial vapor pressures comparable to or higher than water, are transported across hydrophobic membranes with the water vapor.

3.3. Practical Considerations and Further Research

Different ALRs (i.e., HRTs) could be applied to the gas-permeable system operated at semi-continuous mode depending on the TAN concentration required in the effluent. From an agronomic point of view, the three main nutrients (N, P, and K) presented in swine manure in Spanish farms are not balanced in relation to crop needs, occasionally containing an excess of N [33]. In those cases, nitrogen capture through the implementation of gas-permeable membrane technologies could balance swine manure nutrients, enhancing the fertilizing properties of the by-product. Moreover, gas-permeable membranes for capturing N from swine manure can also be combined with other treatment technologies to improve their performance, such as anaerobic digestion process [15,34] or phosphorous recovery [24,28]. In the first case, the use of gas-permeable membrane technology for treating swine manure or other wastes with high ammonia concentration could diminish the ammonia toxicity in anaerobic digestion [35]. With regard to P recovery, the increase in pH together with the reduction of TAN concentration and alkalinity after the recovery of N using gas-permeable membranes could promote P recovery using precipitation processes [28]. More specifically, manure aeration without nitrification causes a pH increase due to OH^- release after bicarbonate destruction. This rise in the pH increases the formation of NH_3 [5]. As shown in Figure 5, alkalinity was consumed due to the TAN removal by the process. In Period I, alkalinity consumption was higher than in Period II. This can be attributed to the higher percentage of TAN removed in Period I, thus consuming higher alkalinity. In this way, the combination of both the increasing of pH and reduction of ammonia and alkalinity using gas-permeable membranes encourage P recovery from swine manure and municipal wastewaters using Ca or Mg compounds [24].

Although the main objective of the semi-continuous gas-permeable system was to remove nitrogen from manure, the treatment also reduced organic matter and solid content (Table 5). During Period I, CODt, TS, and VS removals were 37 $\pm$ 12%, 15 $\pm$ 7%, and 17 $\pm$ 8%, respectively (average values from 10 to 30 d). During Period II, corresponding removal efficiencies were slightly lower: 27 $\pm$ 3%, 6 $\pm$ 4% and 7 $\pm$ 5%, respectively (average values from 35–50 d). Only a very low (<0.1%) amount of the organic matter (CODt) lost from manure was recovered in the acidic solution. Thus, the removal of organic and VS compounds could be attributed to biological degradation processes that take place at room temperature. This organic degradation has been reported in previous assays working with gas-permeable membranes. Dube et al. [5] obtained CODt from negligible to 24% for anaerobically digested effluents after treatment with gas-permeable membranes and low-rate aeration. García-González et al. [15] found consistent CODt removals (60–65%) when recovering ammonia from fresh swine manure using in a gas-permeable membrane system with or without aeration. Differences among studies could be attributed to the different biodegradability of the livestock wastes used (digested vs. fresh manure). COD removal could have consequences from a practical point of view. For instance, the reduction of organic content in treated manure will reduce methane production in a further anaerobic process.

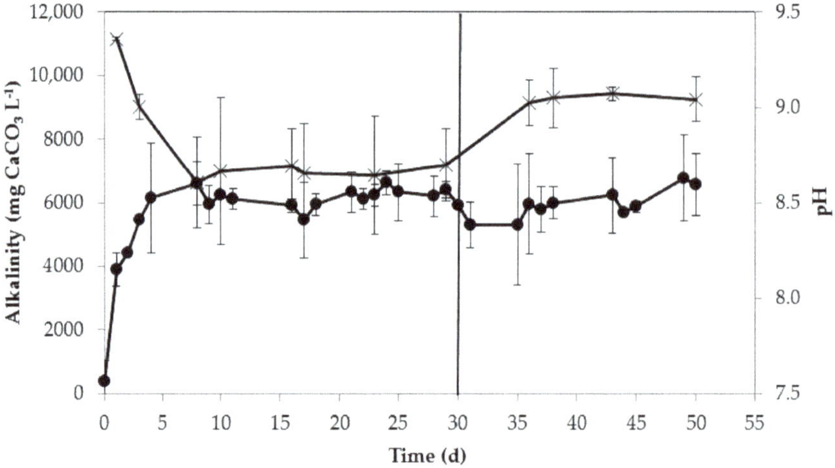

Figure 5. Alkalinity (×) and pH (•) of manure in reactor during the experiment. The error bars represent the standard deviation of duplicate experiments.

Table 5. Average COD, TS, and VS removal efficiencies (± standard deviations) during Period I and II.

	Period I	Period II
COD removal (%)	36.6 ± 11.6	26.5 ± 3.2
TS removal (%)	15.4 ± 7.1	6.2 ± 3.7
VS removal (%)	17. 1 ± 7.5	6.6 ± 5.3

Membrane fouling is an important consideration that determines useful life of the gas-permeable membranes and affects its economic viability [31]. In the present work, the surface of the membrane on the manure side changed color from white to brownish during the 50 days of semi-continuous operation using manure. However, reduction of the rate of TAN recovery over time was not detected. This observation suggests that the membrane soiling did not block membrane pores and did not impact TAN recovery.

The results obtained in the present study are very promising, being the first study (to the best of our knowledge) evaluating gas-membrane technology to recover nitrogen from manure in semi-continuous mode. Nevertheless, further research is needed in order to optimize the process before scaling up this technology. Particularly, in view of the obtained results, the effect of different aeration rates over the pH increase for enhancing N recovery should be evaluated.

4. Conclusions

TAN was successfully removed and recovered from swine manure using gas-permeable membrane system operated at semi-continuous mode. A uniform TAN recovery rate of 27 g m^{-2} d^{-1} was obtained, regardless of the TAN loading rate applied and the manure TAN concentration in reactor. TAN removal reached 79% for Period I (HRT = 7 d) and 56% for Period II (HRT = 5 d), with 90% of recovery by the semi-continuous membrane system in both periods. Simultaneously, ammonia was converted to ammonium sulfate, obtaining a solution of up to 1.9% of N. Osmotic distillation during the recovery process led to the dilution of this acidic solution, reducing the N concentration of the fertilizer product. However, an increase in temperature of 3 °C of the acidic solution relative to the wastewater reduced 34% the osmotic distillation and water dilution of the product.

Author Contributions: The conceptualization of this research was made by B.R. The formal analysis, investigation, and data curation was conducted by B.R. and B.M.-S. Supervision was made by M.C.G.-G. Original draft presentation was done by B.R. Finally, review and editing of the manuscript were prepared by B.R., B.M.-S., M.B.V., and M.C.G.-G.

Funding: This work has been funded by the European Union under the Project Life+ AMMONIA TRAPPING (LIFE15-ENV/ES/000284) "Development of membrane devices to reduce ammonia emissions generated by manure in poultry and pig farms". Cooperation with USDA-ARS Project 6082-13630-001-00D "Improvement of Soil Management Practices and Manure Treatment/Handling System of the Southern Coastal Plains" is acknowledged. Mention of trade names or commercial products in this article is solely for the purpose of providing specific information and does not imply recommendation or endorsement by the USDA.

Conflicts of Interest: The authors declare no conflict of interest.

References

1. Webb, J.; Menzi, H.; Pain, B.F.; Misselbrook, T.H.; Dämmgen, U.; Hendriks, H.; Döhler, H. Managing ammonia emissions from livestock production in Europe. *Environ. Pollut.* **2005**, *135*, 399–406. [CrossRef]
2. European Environment Agency (EEA). Agriculture, Ammonia Emissions Statistics- Data Extracted in June 2015. Available online: http://ec.europa.eu/eurostat/statistics-explained/index.php/Agriculture_-_ammonia_emission_statistics (accessed on 30 June 2017).
3. Wing, S.; Wolf, S. Intensive livestock operations, health, and quality of life among eastern North Carolina residents. *Environ. Health. Perspect.* **2000**, *108*, 233–238. [CrossRef]
4. Directive (EU) 2016/2284 of the European Parlament and of the Council of 14 December 2016 on the reduction of national emissions of certain atmospheric pollutants, amending Directive 2003/35/EC and repeling Directive 2001/81/EC. *Off. J. Eur. Commun.* **2016**, *L344*, 1–31.
5. Dube, P.J.; Vanotti, M.B.; Szogi, A.A.; Garcia-González, M.C. Enhancing recovery of ammonia from swine manure anaerobic digester effluent using gas-permeable membrane technology. *Waste Manag.* **2016**, *49*, 372–377. [CrossRef]
6. Funderburg, E. *Why Are Nitrogen Prices So High?* Agriculture News and Views 2001; The Samuel Roberts Noble Foundation: Ardmore, OK, USA, 2013. Available online: www.noble.org/ag/soils/nitrogenprices/ (accessed on 30 June 2017).
7. Sareer, O.; Mazahar, S.; Khanum Al Akbari, W.M.; Umar, S. Nitrogen pollution, plants and human health. In *Plants, Pollutants and Remediation*; Springer: Dordrecht, The Netherlands, 2016; pp. 47–57.
8. Xie, M.; Shon, H.K.; Gray, S.R.; Elimelech, M. Membrane-based processes for wastewater nutrient recovery: Technology, challenges, and future directions. *Water Res.* **2016**, *89*, 210–221. [CrossRef]
9. Masse, L.; Massé, D.I.; Pellerin, Y.; Dubreuil, J. Osmotic pressure and substrate resistance during the concentration of manure nutrients by reverse osmosis membranes. *J. Membr. Sci.* **2010**, *348*, 28–33. [CrossRef]
10. Bonmatí, A.; Flotats, X. Air stripping of ammonia from pig slurry: Characterization and feasibility as a pre-or post-treatment to mesophilic anaerobic digestion. *Waste Manag.* **2003**, *23*, 261–272. [CrossRef]
11. Milan, Z.; Sánchez, E.; Weiland, P.; de Las Pozas, C.; Borja, R.; Mayari, R.; Rovirosa, N. Ammonia removal from anaerobically treated piggery manure by ion exchange in columns packed with homoionic zeolites. *Chem. Eng. J.* **1997**, *66*, 65–71. [CrossRef]
12. Uludag-Demirer, S.; Demirer, G.N.; Chen, S. Ammonia removal from anaerobically digested dairy manure by struvite precipitation. *Process Biochem.* **2005**, *40*, 3667–3674. [CrossRef]
13. Vanotti, M.B.; Szogi, A.A. Systems and Methods for Reducing Ammonia Emissions from Liquid Effluents and for Recovering the Ammonia. U.S. Patent 9,005,333 B1, 14 April 2015.
14. García-González, M.C.; Vanotti, M.B. Recovery of ammonia from swine manure using gas-permeable membranes: Effect of waste strength and pH. *Waste Manag.* **2015**, *38*, 455–461. [CrossRef]
15. García-González, M.C.; Vanotti, M.B.; Szogi, A.A. Recovery of ammonia from swine manure using gas-permeable membranes: Effect of aeration. *J. Environ. Manag.* **2015**, *152*, 19–26. [CrossRef]
16. Zarebska, A.; Romero Nieto, D.; Chirstensen, K.V.; Fjerbaek Sotoft, L.; Norddahl, B. Ammonium fertilizers production from manure: A critical review. *Crit. Rev. Environ. Sci. Technol.* **2015**, *45*, 1469–1521. [CrossRef]
17. Daguerre-Martini, S.; Vanotti, M.B.; Rodríguez-Pastor, M.; Rosal, A.; Moral, R. Nitrogen recovery from wastewater using gas-permeable membranes: Impact of inorganic carbon content and natural organic matter. *Water Res.* **2018**, *137*, 2010–2210. [CrossRef]

18. Samani Majd, A.M.; Mukhtar, S. Ammonia recovery enhancement using a tubular gar-permeable membrane system in laboratory and field-scale studies. *Trans. ASABE* **2013**, *56*, 1951–1958.
19. Ahn, Y.T.; Hwang, Y.H.; Shin, H.S. Application of PTFE membrane for ammonia removal in a membrane contactor. *Water Sci. Technol.* **2011**, *63*, 2944–2948. [CrossRef]
20. Rothrock, M.J.; Szögi, A.A.; Vanotti, M.B. Recovery of ammonia from poultry litter using gas-permeable membranes. *Trans. ASABE* **2010**, *53*, 1267–1275. [CrossRef]
21. Samani Majd, A.M.; Mukhtar, S. Ammonia diffusion and capture into a tubular gas-permeable membrane using diluted acids. *Trans. ASABE* **2013**, *56*, 1943–1950.
22. García-González, M.C.; Vanotti, M.B.; Szogi, A.A. Recovery of ammonia from anaerobically digested manure using gas-permeable membranes. *Sci. Agric.* **2016**, *73*, 434–438. [CrossRef]
23. Oliveira Filho, J.D.S.; Daguerre-Martini, S.; Vanotti, M.B.; Saez-Tovar, J.; Rosal, A.; Pérez-Murcia, M.D.; Bustamante, M.A.; Moral, R. Recovery of ammonia in raw and co-digested swine manure using gas-permeable membrane technology. *Front. Sustain. Food Syst.* **2018**, *2*, 30. [CrossRef]
24. Vanotti, M.B.; Szogi, A.A.; Dube, P.J. Systems and Methods for Recovering Ammonium and Phosphorous from Liquid Effluents. U.S. Patent 20160347630 A1, 1 December 2016.
25. American Public Health Association. Standard Methods for the Examination of Water, Wastewater APHA. In *American Water Works Association and Water Environment Federation*, 21st ed.; American Public Health Association: Washington, DC, USA, 2005.
26. US Environmental Protection Agency (EPA). *Methods for Chemical Analysis of Water and Waste, EPA/600/4-79/020*; US Environmental Protection Agency: Cincinnati, Ohio, 1983.
27. Hansen, K.H.; Angelidaki, I.; Ahring, B.K. Anaerobic digestion of swine manure. Inhibition by ammonia. *Water Res.* **1998**, *32*, 5–12. [CrossRef]
28. Vanotti, M.B.; Dube, P.J.; Szogi, A.A.; García-González, M.C. Recovery of ammonia and phosphate minerals from swine wastewater using gas-permeable membranes. *Water Res.* **2017**, *112*, 137–146. [CrossRef]
29. Darestani, M.; Haigh, V.; Couperthwaite, S.J.; Millar, G.J.; Nghiem, L.D. Hollow fibre membrane contactors for ammonia recovery: Current status and future development. *J. Environ. Chem. Eng.* **2017**, *5*, 1349–1359. [CrossRef]
30. Wang, G.; Shi, H.; Shen, Z. Influence of osmotic distillation on membrane absorption for the treatment of high strength ammonia wastewater. *J. Environ. Sci.* **2004**, *16*, 651–655.
31. Zarebska, A.; Romero Nieto, D.; Christensen, K.V.; Norddahl, B. Ammonia recovery from agricultural wastes by membrane distillation: Fouling characterization and mechanism. *Water Res.* **2014**, *56*, 1–10. [CrossRef]
32. Cokgor, E.U.; Zengin, G.E.; Tas, D.O.; Oktay, S.; Randall, C.; Orhon, D. Respirometric assessment of primary sludge fermentation product. *J. Environ. Eng.* **2006**, *132*, 68–74. [CrossRef]
33. Antezana, W.; De Blas, C.; García-Rebollar, P.; Rodríguez, C.; Beccaccia, A.; Ferrer, P.; Cerisuelo, A.; Moset, V.; Estellés, F.; Cambra-López, M.; et al. Composition, potential emissions and agriculture value of pig slurry from Spanish commercial farms. *Nutr. Cycl. Agroecosyst.* **2016**, *104*, 159–173. [CrossRef]
34. Lauterböck, B.; Nikolausz, M.; Lv, Z.; Baumgartner, M.; Liebhard, G.; Fuchs, W. Improvement of anaerobic digestion performance by continuous nitrogen removal with a membrane contactor treating substrate rich in ammonia and sulfide. *Bioresour. Technol.* **2014**, *158*, 209–216. [CrossRef]
35. Lauterböck, B.; Ortner, M.; Haider, R.; Fuchs, W. Counteracting ammonia inhibition in anaerobic digestion by removal with a hollow fiber membrane contactor. *Water Res.* **2012**, *46*, 4861–4869. [CrossRef]

environments

MDPI

Article

Effect of the Type of Gas-Permeable Membrane in Ammonia Recovery from Air

María Soto-Herranz [1,*], Mercedes Sánchez-Báscones [1], Juan Manuel Antolín-Rodríguez [1], Diego Conde-Cid [1] and Matias B. Vanotti [2]

1 Department of Agroforestry Sciences, ETSIIAA, University of Valladolid, Avenida de Madrid 44, 34004 Palencia, Spain; msanchez@agro.uva.es (M.S.-B.); juanmanuel.antolin@uva.es (J.M.A.-R.); diego.dy15@gmail.com (D.C.-C.)

2 United States Department of Agriculture (USDA), Agricultural Research Service, Coastal Plains Soil, Water and Plant Research Center, 2611 W. Lucas St., Florence, SC 29501, USA; Matias.Vanotti@ars.usda.gov

* Correspondence: m.sotoh16@gmail.com; Tel.: +34-650-622-390

Received: 7 May 2019; Accepted: 14 June 2019; Published: 16 June 2019

Abstract: Animal production is one of the largest contributors to ammonia emissions. A project, "Ammonia Trapping", was designed to recover gaseous ammonia from animal barns in Spain. Laboratory experiments were conducted to select a type of membrane most suitable for gaseous ammonia trapping. Three types of gas-permeable membranes (GPM), all made of expanded polytetrafluoroethylene (ePTFE), but with different diameter (3.0 to 8.6 mm), polymer density (0.45 to 1.09), air permeability (2 to 40 $L \cdot min^{-1} \cdot cm^2$), and porosity (5.6 to 21.8%) were evaluated for their effectiveness to recover gas phase ammonia. The ammonia evolved from a synthetic solution (NH_4Cl + $NaHCO_3$ + allylthiourea), and an acidic solution (1 N H_2SO_4) was used as the ammonia trapping solution. Replicated tests were performed simultaneously during a period of 7 days with a constant flow of acidic solution circulating through the lumen of the tubular membrane. The ammonia recovery yields were higher with the use of membranes of greater diameter and corresponding surface area, but they were not affected by the large differences in material density, porosity, air permeability, and wall thickness in the range evaluated. A higher fluid velocity of the acidic solution significantly increased—approximately 3 times—the mass NH_3–N recovered per unit of membrane surface area and time (N-flux), from 1.7 to 5.8 mg $N \cdot cm^{-2} \cdot d^{-1}$. Therefore, to optimize the effectiveness of GPM system to capture gaseous ammonia, the appropriate velocity of the circulating acidic solution should be an important design consideration.

Keywords: ammonia recovery; ammonia capture; air pollution; gas-permeable membrane; ammonium sulfate

1. Introduction

Animal production is one of the largest contributors to ammonia emissions (NH_3) [1] due to poor waste management. Ammonia is implicated in particulate formation (PM 2.5) with adverse effects on human health [2]. Ammonia also contributes to ecosystem degradation when it is deposited on land or water [3] with corresponding soil acidification and eutrophication of surface water bodies [4].

In 2016, the agricultural sector of the EU-28 was responsible for 92% of the total ammonia emissions in the region because of the volatilization of livestock excreta [5]. In Spain, according to the National Emissions Inventory (1990–2015), agricultural activities produced 96% of the ammonia emissions. In 2014 and 2015, the National Emission Ceilings for the NH_3 (353 $kt \cdot year^{-1}$) were exceeded by 7% [6]. According to EU Directive 2016/2284/EU [7], Spain must reduce the NH_3 emission ceiling by 3% during the period 2020–2029 and by 16% by the year 2030.

Mechanical ventilation is considered a basic control method to eliminate gaseous ammonia from the inside of livestock production barns [8] to ensure the health of workers and animals [9] and the animal production performance [10].

The application of gas-permeable membranes (GPM) for capturing ammonia has been tested, especially in liquid [11,12]. Methods designed for the capture and recovery of N as a resource are the most optimal [13–15]. Conservation and recovery of N are important in agriculture due to the high cost of commercial ammonia fertilizers [16]. In this way, it would contribute positively from both an environmental point of view (decreased ammonia emissions to the atmosphere) and an economic point of view (recovered ammonium to replace commercial fertilizers of nitrogen source). Furthermore, the advantages of gas-permeable membrane technology over other N recovery technologies are, among others, that it does not require the use of additives [17], and it has low energy consumption in relation to other methods of ammonia recovery [18].

The GPM process consists in the flux of ammonia gas through the microporous hydrophobic membrane by diffusion. This ammonia is captured in an acidic solution circulating inside the membrane. As shown in Equation (1), once in contact with the acidic solution, the NH_3 gas combines with the free protons of the acid to form non-volatile ammonium ions (NH_4^+). When sulfuric acid is used in the process, the product is ammonium sulfate. Sulfuric acid is generally used as a source of acid to capture ammonia because of its lowest cost among inorganic acids. However, the process is also effective using other inorganic acids (nitric, phosphoric), organic acids (citric, lactic), and their precursors [19]. Additionally, ammonium sulfate (AS) may have some potential agronomic and environmental benefits compared with ammonium nitrate (AN) by creating a more acidic root rhizosphere that increases the availability of soil P, and by reducing denitrification in soil and N_2O greenhouse gas emissions [20]. Therefore, it can be an adequate substitute of mineral fertilizers as a nitrogen source and valuable fertilizer.

$$2NH_3 + H_2SO_4 \rightarrow (NH_4)_2SO_4 \quad (1)$$

For gas separation and recovery, organic hydrophobic gas-permeable membranes (GPM), especially expanded polytetrafluoroethylene (ePTFE), are preferred due to lower transference resistance, hydrophobic characteristics, organic resistance, and chemical stability with acidic solutions [21,22].

The final mass of NH_3 captured in the acidic solution depends on the concentration of NH_3 gas in the atmosphere, which depends on the pH and the TAN (total ammonia nitrogen, NH_3 + NH_4^+) concentration of the emitting solution [23], pH of the acidic solution [24], and acidic solution flow rate [25].

In animal manures, NH_3 and NH_4^+ are in equilibrium depending on the pH and the temperature. The ammonia dissolves at the source and/or is emitted. At pH below 7, little of the ammonia is undissociated and present as dissolved gas in liquid mixtures [26], for example, only 0.36% at pH 6.8 and temperature 25 °C [27]. At higher pH, a higher concentration of the undissociated, free NH_3 is instantly produced (i.e., 26.4, 78.2, and 97.3% at pH 8.8, 9.8, and 10.8, respectively). These conditions favor NH_3 permeation through the membrane where an acidic solution circulates [28,29]. With a pH < 2, the acidic solution dissolves the NH_3, transforming it into an ammonium salt [11].

Most ammonia capture applications with gas-permeable tubular membranes have been performed in the liquid state (effluents). An EU project, "Ammonia Trapping (AT)", was designed to recover gaseous ammonia from animal barns in Spain. The main objective of the AT project was to reduce NH_3 gas emissions from the atmosphere of swine and poultry farms by using gas-permeable ePTFE membranes and capturing the N directly from the air. Targets in AT project were a reduction in the NH_3 concentration > 70%, and flux rates of ammonia trapping of 1.3 g N m^{-2} d^{-1}. The goal of this study was to determine the efficiency of the different gas-permeable tubular membranes to capture ammonia from the air. The results of this laboratory study helped in selection of the materials before a larger on-farm pilot evaluation, especially given the difference in costs of these membranes. To avoid variations in ammonia emissions, among treatments each experiment used the same synthetic

N emitting solution. In addition, the pH of the acidic solution used for ammonia capture was kept below 2.

2. Materials and Methods

2.1. Experimental Design

Airtight chambers were used to recover gaseous ammonia using the method of Szogi et al. [19]. The experiment used three chambers (volume = 25 L), one for testing each type of gas-permeable membrane (Figure 1). The lids of the chambers were sealed. A tank (volume = 11 L) containing 1 L of a synthetic N emitting solution was placed inside of each chamber. Tubular gas-permeable membranes were suspended in the air in the chambers. The membranes were connected to an acidic solution reservoir that contained 1 L of an acidic N capturing solution (1 *N* H_2SO_4). Peristaltic pumps (Minipuls 2, Gylson, USA or Perimax 12, Spetec, Germany) recirculated the acidic solution in a closed loop [16] between the inside of the tubular membranes and the acidic solution reservoir.

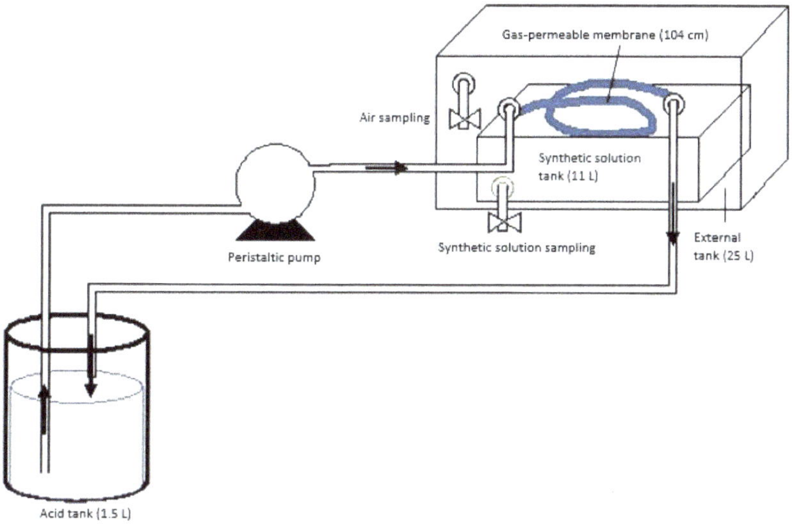

Figure 1. Diagram of the process of ammonia uptake by the gas-permeable membrane in a closed loop.

The membranes were made of expanded polytetrafluoroethylene (ePTFE), but with different characteristics: ZM, ZM4, FZM (ZEUS Industrial Products Inc., Orangeburg, SC, USA), and PM (PRODYSOL Company) (Table 1).

Table 1. Characteristics of the membranes.

Membrane Characteristics	ZM	FZM	PM	PM	ZM4
		Experiment 1		Experiment 2	
Length (cm)	104.0	70.0	104.0	46.3	100.0
Outer diameter (mm)	8.6	3.0	8.6	8.6	4.1
Width of the wall (mm)	0.8	1.0	1.2	1.2	0.6
Average pore size length (μm) *	27.6 ± 8.3	5.8 ± 0.8	14.7 ± 2.3	14.7 ± 2.3	-
Average pore size width (μm) *	7.6 ± 0.9	0.7 ± 0.1	5.5 ± 0.6	5.5 ± 0.6	-
Polymer density (g/cm^3)	0.45	1.09	0.95	0.95	0.95
Absorption surface (cm^2)	282.3	66.0	282.3	125.7	125.7

The membrane abbreviations are ZM: Zeus membrane (8.6 mm outer diameter), FZM: Zeus membrane (3.0 mm outer diameter), PM: Prodysol membrane (8.6 mm outer diameter), ZM4: Zeus membrane (4.1 mm outer diameter).
* The membrane pores of the expanded polytetrafluoroethylene (ePTFE) membranes were elongated (Figure 2). Pore sizes were reported for both length and width by measuring 10 pores in the SEM.

Two experiments were conducted. In experiment 1, three conditions were evaluated. The same flow rate of the acidic solution (1.25 L h^{-1}) was applied to the three treatments. Two treatments used membranes with contrasting characteristics (ZM and PM) and with equal surface area (282.3 cm^2). The third treatment used a membrane (FZM) with smaller diameter and lower surface area (66 cm^2) (Table 1).

Experiment 2 was conducted to verify the effect of the fluid velocity on ammonia flux. Four conditions were evaluated. The four treatments used the same membrane surface (125.7 cm^2).

Two membrane types with different diameter were tested (PM and ZM4), and each membrane type received two acid flow rates (0.83 and 1.5 L h^{-1}).

A synthetic solution was used in both experiments as the source of NH_3 emission (instead of organic waste). In the first experiment, the N emitting solution contained 59.4 g L^{-1} NH_4Cl, 108.5 g L^{-1} $NaHCO_3$, and 10 mg·L^{-1} N-allylthiourea, and in the second experiment, the N emitting solution contained 24.6 g L^{-1} NH_4Cl, 43.2 g L^{-1} $NaHCO_3$, and 10 mg·L^{-1} N-allylthiourea. N-allylthiourea (98%) was added as a nitrification inhibitor, following the strategies presented in other assays [22,30].

Two repetitions were made with each treatment tested. Samples (7.5 mL) of the N emitting synthetic solution were collected every two days and samples (5 mL) of the N trapping acidic solution were collected every day. The room temperature was constant (20.0 ± 1.0 °C).

2.2. Methodology for Analyses

Temperature (°C), pH, and TAN concentration (mg·L^{-1}) were monitored in the acidic solutions and the N emitting solution. The weight variations were controlled in the acidic solution reservoir, taking into account the 5 mL of sample extracted.

The control of pH was realized in the N trapping acidic solution and the N emitting synthetic solutions: pH of the acidic solution was maintained at < 2 and the pH of the synthetic solution at > 8 [11,12,30]. pH modifications were not required because these conditions were not reached.

The pH was measured with a Crison GLP22 pH meter (Crison Instruments S.A., Barcelona, Spain). The ammonium analysis was performed with distillation (UDK 140 automatic steam distillation unit, Velp scientific), capture of distillate in borate buffer, and subsequent titration with 0.2 N HCl [31].

The internal surface morphology of the membranes (Figure 2) was analyzed by scanning electron microscopy (SEM) in the Advanced Microscopy Unit of the University of Valladolid. The SEM images were obtained using a FEI QUANTA 200F device (FEI Co, USA). The pore size distribution (pores/m^2), porosity, and water and air permeability were measured using porosimetry equipment (Coulter Porometer II) [32]. The surface sizes of the FPM and ZM4 membranes were not suitable for porosimetry and permeability analyses, so these measurements were obtained for ZM and PM membranes only.

Data were analyzed by means and standard deviation. Linear regression analyses were used to quantify changes of weight of the acidic solution and N capture rates. Data related to N mass removed, N mass recovered, and N flux were subjected to ANOVA (SAS Institute, 2008) [33].

2.3. Mass Flow Calculation

Mass flow (J) of NH_3–N or N flux (mg N·cm^{-2}·d^{-1}) from the air into the acidic solution was calculated based on the N mass captured per day and the surface area of the GPM tubing using Equation (2), where C is the concentration of NH_4–N in the acidic solution (mg·L^{-1}), V is the volume of the acidic solution (L), S the contact surface of the membrane (cm^2), and t the time (d).

$$J = (C \times V)/(S \times t) \quad (2)$$

3. Results and Discussion

3.1. Characterization of Membranes

Measurements of the number of pores, porosity, water and air permeability, and MFP (mean flow pore) for the membranes ZM and PM are shown in Table 2.

Table 2. Values of number of pores, porosity, water permeability, and MFP (mean flow pore) of ZM and PM membranes in experiment 1.

Type of Membrane	N° pores (pores/m^2)	Porosity (%)	Water permeability (L·min^{-1})	Air permeability (L·min^{-1}·cm^2)	MFP (μm)
ZM	$1.2 \cdot \times 10^{11} \pm 4.1 \cdot \times 10^{10}$	21.8 ± 3.2	$2.5 \cdot \times 10^{-7} \pm 6.8 \cdot \times 10^{-9}$	10–25–40 [a]	1.7 ± 0.1
PM	$5.2 \cdot \times 10^{10} \pm 1.4 \cdot \times 10^{10}$	5.6 ± 0.9	$1.3 \cdot \times 10^{-7} \pm 2.0 \cdot \times 10^{-8}$	2–5–10 [a]	1.2 ± 0.1

([a]) The air permeability was estimated at three pressures (1, 2, and 3 bars of pressure).

Compared with the PM membrane, the ZM membrane had a lower density (0.45), a higher number and size of pores, and a higher porosity and permeability (Table 2). This result can also be verified by the SEM images in Figure 2, which indicate that the pore size was greater in ZM, followed by PM, and finally FZM.

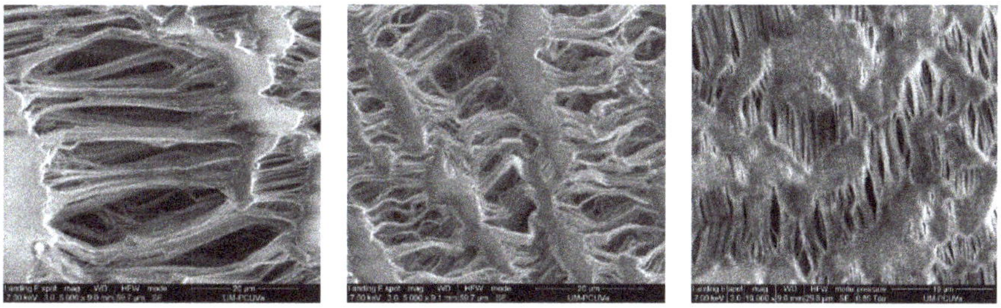

(a) (b) (c)

Figure 2. Scanning electron microscopy (SEM) images for the ZM (**a**), PM (**b**), and FZM (**c**) ePTFE membranes showing typical elongated pores of different sizes structures. The images correspond to the inner surface of the tubular membranes. Images A and B were taken with 5000× magnification and the scale bar is equivalent to 20 μm in length. Image C were taken with 10,000× magnification and the scale bar is equivalent to 10 μm in length.

3.2. Variation of the Weight of the Acidic Solution

The acidic solution decreased in weight in the three GPM systems (Figure 3). Total weight losses of the acidic solution at the end of the experiment were 11 ± 2% for ZM (R2 = 0.89), 10 ± 4% for PM (R2 = 0.95), and 5 ± 1% for FZM (R2 = 0.99). Weight losses in all cases were related with an evaporation process as leaks were not observed. The rate of water weight loss (g·d^{-1}) for each type of membrane was: 16.1 ± 6.1 g·d^{-1} for PM, 17.8 ± 2.9 g·d^{-1} for ZM, and 7.5 ± 1.2 g·d^{-1} for FZM. The hydrophobic nature of the membrane prevents the penetration of the acidic solution into the membrane pores, creating a liquid/vapor interface at each pore entrance. If a vapor partial pressure difference across the membrane is established, vapor transport across the membrane takes place [34].

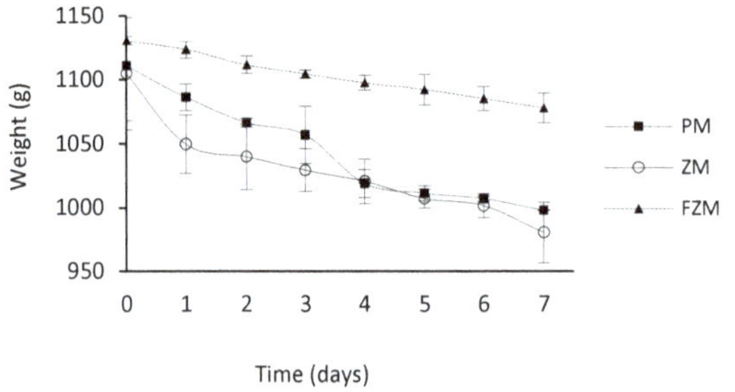

Figure 3. Weight loss of the acidic solution for each membrane type (experiment 1).

Majd et al. [23] also observed volume losses in acid traps due to evaporation, with values between 1 to 2 $mL \cdot d^{-1}$ in suspended systems, with an acid volume of 190 mL and a flow rate 3 times lower than that used in this experiment.

The rate of weight loss of the acid was not affected by membrane density, porosity, and permeability; however, it was affected by surface area. Weight loss was higher in the two membranes (ZM and PM) with the larger diameter and surface area, even though they had different porosity and density, and the weight loss was lower in the FZM membrane with smaller diameter and surface area. Therefore, the greater surface area resulted in higher vapor transport across the membranes and acid weight loss.

3.3. *Process pH in the N Capturing Acidic Solution and N Emitting Synthetic Solution*

The pH values reached in the acidic solution for each type of membrane were 0.4 ± 0.1 for PM, 0.4 ± 0.1 for ZM, and 0.4 ± 0.1 for FZM. In all cases, the pH values reached in the acidic solution remained below 2, indicating that enough H^+ ions were available to react continuously with NH_3 [25], forming an ammonium salt.

The initial pH values in the N-emitting synthetic solution for each type of membrane were 8.74 ± 0.06 for PM, 8.76 ± 0.01 for ZM, and 8.78 ± 0.13 for FZM. Corresponding final pH values at day 7 were 8.27 ± 0.04, 8.45 ± 0.01, and 8.63 ± 0.06. In all cases, the pH of the synthetic solution was maintained above 8, which favored the emission of free NH_3 [30].

3.4. *Effect of the Type of Membrane on Ammonia Capture*

The total NH_3–N mass emitted by the synthetic solution was similar in the three membrane systems: 5381 ± 451 mg N for PM membrane, 5260 ± 514 mg N for ZM, and 4764 ± 606 mg N for FZM (Table 3).

Table 3. Mass of NH_3–N removal, NH_3–N recovered by gas-permeable membranes, and N-flux with varied polymer density, surface area, and acidic solution velocity (experiment 1).

Type of Membrane	e-PTFE Density ($g\ cm^{-3}$)	i.d. [1] (mm)	Acidic Solution Velocity [2] ($cm\ min^{-1}$)	Surface Area (cm^2)	NH_3–N Mass Removed [3] (mg)	NH_3–N Mass Recovered (mg)	N flux ($mg\ N \cdot cm^{-2} \cdot d^{-1}$)
PM	0.95	6.2	69	282.3	5381 a [4]	3407 a	1.7 b
ZM	0.45	7.0	54	282.3	5260 a	3628 a	1.8 b
FZM	1.09	1.0	2654	66.0	4764 a	2661 b	5.8 a

[1] i.d. = inner diameter of the tubular membrane; [2] acidic solution velocity inside the tubular membrane. Flow rate was constant across membranes (1.25 L/h). Reynolds numbers were 73, 64, and 415 for PM, ZM, and FZM; [3] N mass removed from the N emitting synthetic solution; [4] values in a column followed by the same letter are not significantly different ($p \leq 0.05$).

Corresponding percent N removals were 46 ± 4%, 45 ± 4%, and 41 ± 5%. Similarly, no differences were observed in the total mass of NH_3–N present in the synthetic solutions at the end of the experiment (Figure 4).

The ammonia emission rate of the synthetic solution varied with time. There was a higher emission rate on the first day and a lower and almost constant emission rate in later days. For example, rates of emission the first day were 4138 ± 47 mg NH_3–N·d^{-1} for PM, 3555 ± 433 mg NH_3–N·d^{-1} for ZM, and 3342 ± 463 mg NH_3–N·d^{-1} for FZM, and afterwards the rates of emission were 207 ± 83 for PM, 284 ± 158 for ZM, and 237 ± 24 mg NH_3–N·d^{-1} for FZM. This emission behavior was also observed by Rothrock et al. [23] who noted that in the first 7 days, the concentration of NH_4–N present in the synthetic source solution decreased faster, from 500 mg to 300 mg approximately. In contrast, from days 7 to 21, the concentration only decreased from 300 mg to 200 mg. The high recovery observed on the first day could be due to the high concentration of ammonium in the synthetic solutions. This generates a high concentration of ammonia in the gas phase. After the first day, a significant percentage of ammonium had been eliminated and, therefore, the driving force for transport decreased.

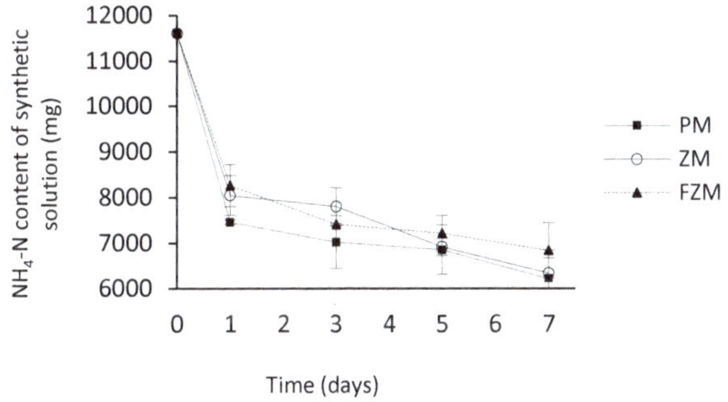

Figure 4. Mass of NH_3–N remaining in the N emitting synthetic solution for three different types of ePTFE tubular membranes (experiment 1). Data points are means ± s.d. of duplicate experiments.

The masses of NH_3–N recovered in the acidic N trapping solution were 3628 ± 27 mg, 3407 ± 49 mg, and 2661 ± 307 mg for ZM, PM, and FZM, respectively. At similar emission and capture conditions, the NH_3–N mass recovered by FZM was lower due to a lower surface area compared to the membranes ZM and PM. The surface area was 4.2 times higher for PM and ZM compared to FZM. Surprisingly, the mass of NH_3–N recovered was not affected by large differences in material density (0.45 to 0.95 g/cm^3) between PM and ZM (Table 3), or by differences in porosity (5.6 to 21.8%), air permeability (2 to 10 L·min^{-1} cm^{-1} at 1 bar pressure), and wall thickness (0.8 to 1.2 mm). This was surprising because it is logical to think that higher NH_3 capture should be obtained with higher membrane porosity and air permeability, and with smaller wall thickness [35]. However, in the range tested in the experiment, these characteristics did not affect mass of NH_3 recovered by the membranes.

In all membrane systems, NH_3–N accumulation in the acidic solution during the 7-day experimental period was linear (Figure 5). Capture rates (mg NH_3–N d^{-1}) were calculated based on the slope of the linear regressions; they were higher with PM and ZM (487 ± 71 and 518 ± 4 mg NH_3–N d^{-1}, respectively) with larger diameter and surface area, compared to FZM (380 ± 44 mg NH_3–N d^{-1}) with smallest diameter and surface area. The acidic solution had a similar composition of 0.3 ± 0.1% of nitrogen and 0.4 ± 0.1% of sulphur.

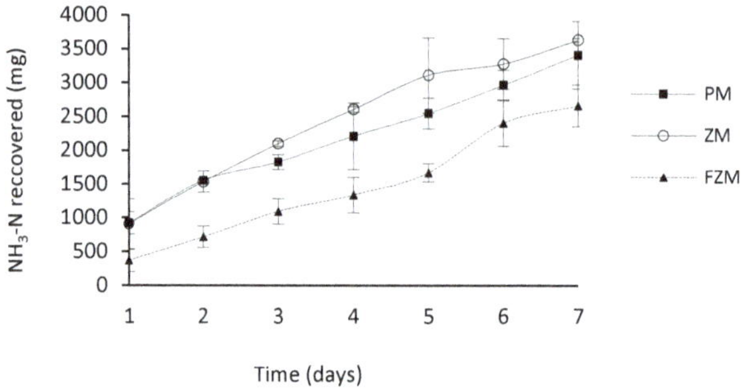

Figure 5. Mass of NH_3–N recovered in the acidic solution for three different types of ePTFE tubular membranes (experiment 1). Data points are means ± s.d. of duplicate experiments.

The NH_3 recovery (%) for each type of membrane was calculated based on the relationship between the NH_3–N mass recovered (final content of NH_3–N in the trapping solution) and the NH_3–N mass removed (difference between the initial and final content of NH_3–N in synthetic solution). Percent recoveries were not different ($p \leq 0.05$): PM = 63%, ZM membrane = 69%, and FZM = 57%. The percent recoveries were not quantitative (100%) probably because the rapid release of NH_3 in the first day of the experiment exceeded the capacity of the membrane. Other authors such as Rothrock et al. [23] obtained similar results than under conditions of an NH_3 emission flush. They achieved recoveries of NH_3–N of 67.7%, 73.6%, and 76.2% with hydrated lime addition treatments of 0.4 w/v, 2 w/v, and 4 w/v to 300 g of poultry litter. Therefore, design of the membrane manifolds should consider possible situations of rapid release that may occur in filed situations such as disinfection of manure with alkali compounds.

On the other hand, when the NH_3–N capture is expressed on a surface area basis (N-flux, Table 3), the results show additional insight on the best operating conditions for the membranes. The N-flux obtained in the membrane FZM with the smaller diameter (5.8 ± 0.7 mg $N \cdot cm^{-2} \cdot d^{-1}$) was significantly higher—approximately 3 times higher—compared with the N-flux obtained with the larger diameter membranes (1.8 ± 0.0 and 1.7 ± 0.2 mg $N \cdot cm^{-2} \cdot d^{-1}$ with ZM and PM, respectively). Rothrock et al. [23] also observed higher N-fluxes in membranes with a smaller diameter (1.37 $g \cdot m^{-2} \cdot d^{-1}$ N-flux for a membrane i.d. of 4.0 mm and acid flow 70–80 mL d^{-1} and 0.7 $g \cdot m^{-2} \cdot d^{-1}$ N-flux 0.70 for a membrane i.d. of 8.8 mm and same flow). Majd and Mukhtar [25] observed an N-flux of 0.2 $g \cdot m^{-2} \cdot d^{-1}$ in a suspended membrane system. However, higher ammonia fluxes have been obtained when the membranes were directly submerged in the liquid (liquid–liquid) instead of being suspended in the air (air–liquid). For example, Daguerre et al. [36] obtained N-fluxes of 7.1 to 8.9 $g \cdot m^{-2} \cdot d^{-1}$ placing the membrane manifold in liquid swine manure (4940 mg NH_4–N L^{-1}), and Fillingham et al. [37] obtained N-fluxes up to 51.0 $g \cdot m^{-2} \cdot d^{-1}$ using synthetic wastewaters containing 6130 NH_4–N L^{-1} and NaOH to pH 8.5.

In this study (experiment 1), the same recirculation flow of the acidic solution (1.25 L h^{-1}) was used with the three membranes with outside diameters ranging from 3 to 8.6 mm (inner diameters 1 to 7). As a result, the smaller diameter resulted in a higher fluid velocity inside the membrane (2653.9 cm min^{-1}) compared to the higher diameter membranes (54.2 to 69.0 cm min^{-1}) (Table 3) and a more frequent renovation of the acidic solution in the submerged membrane manifold.

The Reynolds number (Re) is used in fluid dynamics to describe the character of the flow (flow is laminar when Re < 2300 and viscous forces are dominant characterized by smooth fluid motion, and flow is turbulent when Re > 3000 and it is dominated by inertial forces and vortices). Although the fluid flow was laminar in all three cases (Re 64 to 415, Table 3), the higher fluid velocity and Re in FZM resulted in a higher N-flux. Therefore, to optimize the effectiveness of the ePTFE membranes to capture gaseous ammonia, the fluid velocity should be an important design consideration because this study showed that the efficiency can be increased 3 times with changes in acidic solution velocity.

A second experiment was done to further evaluate the positive effect of fluid velocity on N-flux. The study used two recirculation rates (0.83 and 1.5 L h^{-1}), and two membrane types with different diameters (id 2.9 and 6.2 mm) but the same surface area (125.5 cm^2) (Table 4). As a result, the fluid velocity inside the membranes gradually increased in the range of 49 to 315 cm min^{-1} and Re varied from 49 to 155. These modest differences in fluid velocity and Re (within laminar flow) significantly affected both the mass of NH_3–N recovered in the acidic N trapping solution, and the N-flux (recovery per surface area). Figure 6 shows the relationship between N flux vs Re obtained using combined data from all seven treatments in experiments 1 and 2. It confirms that velocity of the circulating acidic solution should be an important design consideration to optimize the effectiveness of GPM system to capture gaseous ammonia.

Table 4. Effect of acidic solution velocity on NH_3–N recovered and N-flux of gas–permeable membranes (experiment 2).

Type of Membrane	e-PTFE Density (g cm^3)	i.d. [1] (mm)	Surface Area (cm^2)	Acidic Solution Flow Rate (L h^{-1})	Acidic Solution Velocity [2] (cm/min)	Reynolds Number [3]	NH_3–N Mass Recovered (mg)	N Flux (mg N·cm^{-2}·d^{-1})
PM	0.95	6.2	125.7	0.83	46	49	3162 bc [4]	1.8 bc
PM	0.95	6.2	125.7	1.25	69	73	2780 c	1.6 c
ZM4	0.95	2.9	125.7	0.83	210	104	3686 ab	2.1 b
ZM4	0.95	2.9	125.7	1.25	315	155	4444 a	2.5 a

[1] i.d. = inner diameter of the tubular membrane; [2] acidic solution velocity inside the tubular membrane; [3] Reynolds number (Re) = v.l/ν, where v = velocity of the fluid (m/s), l = tube i.d. (m), and ν = kinematic viscosity of the liquid at 20 °C (9.79 × 10^{-7} m^2/s); [4] values in a column followed by the same letter are not significantly different ($p \leq 0.05$).

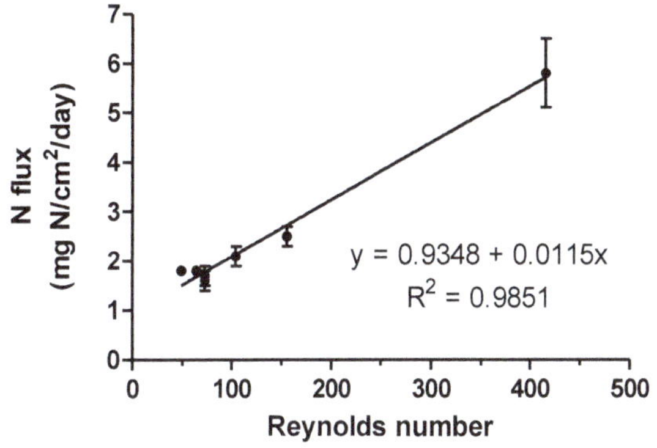

Figure 6. Ammonia flux (mg N per cm^2 of membrane surface per day) through the gas-permeable membrane as affected by changes in Reynolds number (combined data from experiments 1 and 2). Data points are means ± s.d. of duplicate experiments.

4. Conclusions

Gas-permeable membranes (GPM) made of ePTFE were effective for the recovery of gaseous NH_3 using a closed-loop system. A pH < 2 in the circulating acidic solution and a pH > 8 of the synthetic emitting solution were favorable for the process. At similar emission and capture conditions, the mass of NH_3–N recovered by tubular GPMs was significantly increased by surface area, which was related to differences in the membrane diameter tested. However, the mass of NH_3–N recovered was not affected by large differences in GPM material density (0.45 to 0.95 g/cm^3), porosity (5.6 to 21.8%), air permeability (2 to 10 L·min^{-1} cm^{-1} at 1 bar pressure), and wall thickness (0.8 to 1.2 mm). A higher fluid velocity of the acidic solution significantly increased (approximately 3 times) the N-flux (mass N recovered per unit of surface area and time). Therefore, to optimize the effectiveness of the GPM system to capture gaseous ammonia, the fluid velocity is an important design consideration because this study showed that the efficiency can be increased 3 times with changes in acidic solution velocity.

Author Contributions: The conceptualization of this research was made by M.S.-H. and D.C.-C. The formal analysis, investigation, and data curation was conducted by M.S.-H., D.C.-C., and J.M.A.-R. Supervision was made by J.M.A.-R. and M.S.-B. Original draft was done by M.S.-H. Finally, review and editing of the manuscript were prepared by M.S.-H., J.M.A.-R., M.B.V., and M.S.-B.

Funding: The authors gratefully acknowledge funding by the European Union under the Project Life + "Ammonia Trapping" (LIFE15-ENV/ES/000284) "Development of membrane devices to reduce ammonia emissions generated by manure in poultry and pig farms". Mention of trade names or commercial products in this article is solely for the purpose of providing specific information and does not imply recommendation or endorsement by the USDA.

Conflicts of Interest: The authors declare no conflict of interest.

References

1. Beusen, A.; Bouwman, A.; Heuberger, P.; Van Drecht, G.; Van Der Hoek, K. Bottom-up uncertainty estimates of global ammonia emissions from global agricultural production systems. *Atmos. Environ.* **2008**, *42*, 6067–6077. [CrossRef]
2. Erisman, J.W.; Bleeker, A.; Hensen, A.; Vermeulen, A. Agricultural air quality in Europe and the future perspectives. *Atmos. Environ.* **2008**, *42*, 3209–3217. [CrossRef]
3. Sutton, M.A.; Oenema, O.; Erisman, J.W.; Leip, A.; van Grinsven, H.; Winiwarter, W. Too much of a good thing? *Nature* **2011**, *472*, 159–161. [CrossRef] [PubMed]
4. Bouwman, A.F.; Van Vuuren, D.P.; Derwent, R.G.; Posch, M. A global analysis of acidification and eutrophication of terrestrial ecosystems. *Water Air Soil Pollut.* **2002**, *141*, 349–382. [CrossRef]
5. EEA (European Environment Agency). *European Union Emission Inventory Report 1990–2016 under the UNECE Convention on Long-Range Transboundary Air Pollution (LRTAP)*; ISNN: Cambridge, UK, 2018; pp. 1977–8449.
6. MAPAMA (Ministerio de Agricultura y Pesca, Alimentación y Medio Ambiente). *Inventario de Emisiones de España Emisiones de Contaminantes en el Marco de la Directiva de Techos Nacionales de Emisión Serie 1990–2015*; MAPAMA: Madrid, Spain, 2017. Available online: http://www.mapama.gob.es/es/calidad-y-evaluacion-ambiental/temas/sistema-espanol-de-inventario-sei-/documentoresumeninventariotechosespana-serie1990-2015_tcm30-378885.pdf (accessed on 5 March 2019).
7. EU (European Union). *Directive (EU) 2016/2284 of the European Parliament and of the Council of 14 December 2016 on the Reduction of National Emissions of Certain Atmospheric Pollutants, Amending Directive 2003/35/EC and Repealing Directive 2001/81/EC*; EU: Brussels, Belgium, 2016.
8. Cho, M.S.; Ko, H.J.; Kim, D.; Kim, K.Y. On-site application of air cleaner emitting plasma ion to reduce airborne contaminants in pig building. *Atmos. Environ.* **2012**, *63*, 276–281. [CrossRef]
9. Koerkamp, P.W.G.; Metz, J.H.M.; Uenk, G.H.; Phillips, V.R.; Holden, M.R.; Sneath, R.W.; Short, J.L.; White, R.P.; Hartung, J.; Seedorf, J.; et al. Concentration and emission of ammonia in livestock buildings in Northern Europe. *J. Agric. Eng. Res.* **1998**, *70*, 79–95. [CrossRef]
10. Schauberger, G.; Piringer, M.; Heber, A.J. Odour emission scenarios for fattening pigs as input for dispersion models: A step from an annual mean Value to time series. *Agric. Ecosyst. Environ.* **2014**, *193*, 108–116. [CrossRef]
11. García-González, M.C.; Vanotti, M.B. Recovery of ammonia from swine manure using gas-permeable membranes: Effect of waste strength and pH. *Waste Manag.* **2015**, *38*, 455–461. [CrossRef]
12. Vanotti, M.B.; Szogi, A.A. Systems and Methods for Reducing Ammonia Emissions from Liquid Effluents and for Recovering Ammonia. U.S. Patent 9,005,333 B1, 29 October 2015.
13. Vecino, X.; Reig, M.; Bhushan, B.; Gibert, O.; Valderrama, C.; Cortina, J.L. Liquid fertilizer production by ammonia recovery from treated ammonia-rich regenerated streams using liquid-liquid membrane contactors. *Chem. Eng. J.* **2019**, *360*, 890–899. [CrossRef]
14. Adam, M.R.; Othman, M.H.; Samah, R.A.; Puteh, M.H.; Ismail, A.F.; Mustafa, A.; Rahman, M.A.; Jaafar, J. Current trends and future prospects of ammonia removal in wastewater: A comprehensive review on adsorptive membrane development. *Sep. Purif. Technol.* **2019**, *213*, 114–132. [CrossRef]
15. Zhang, C.; Ma, J.; He, D.; Waite, T.D. Capacitive Membrane Stripping for Ammonia Recovery (CapAmm) from Dilute Wastewaters. *Environ. Sci. Technol.* **2018**, *5*, 43–49. [CrossRef]
16. Vanotti, M.B.; Szogi, A.A. Use of Gas-Permeable Membranes for the Removal and Recovery of Ammonia from High Strength Livestock Wastewater. *Proc. Water Environ. Fed.* **2011**, *2011*, 659–667. [CrossRef]

17. Nelson, N.O.; Mikkelsen, R.L.; Hesterberg, D.L. Struvite formation to remove phosphorus from anaerobic swine lagoon effluent. In *Animal, Agricultural and Food Processing Wastes: Proceedings of the Eighth International Symposium*; Moore, J.A., Ed.; American Society of Agricultural Engineers: St. Joseph, MI, USA, 2000; pp. 18–26.
18. Zarebska, A.; Romero-Nieto, D.; Christensen, K.V.; Fjerbæk Søtoft, L.; Norddahl, B. Ammonium fertilizers production from manure: A critical review. *Crit. Rev. Environ. Sci. Technol.* **2015**, *45*, 1469–1521. [CrossRef]
19. Szogi, A.A.; Vanotti, M.B.; Rothrock, M.J. Gaseous ammonia removal system. US Patent 8,906,332 B2, 9 December 2014.
20. Chien, S.H.; Gearhart, M.M.; Villagarcía, S. Comparison of Ammonium Sulfate with Other Nitrogen and Sulfur Fertilizers in Increasing Crop Production and Minimizing Environmental Impact: A Review. *Soil Sci.* **2011**, *176*, 327–335. [CrossRef]
21. Guo, Y.; Chen, J.; Hao, X.; Zhang, J.; Feng, X.; Zhang, H. A novel process for preparing expanded Polytetrafluoroethylene (ePTFE) micro-porous membrane through ePTFE/ePTFE co-stretching technique. *J. Mater. Sci.* **2007**, *42*, 2081–2085. [CrossRef]
22. Rothrock, M.J.; Szögi, A.A.; Vanotti, M.B. Recovery of ammonia from poultry litter using flat gas permeable membranes. *Waste Manage.* **2013**, *33*, 1531–1538. [CrossRef] [PubMed]
23. Rothrock, M.J.; Szögi, A.A.; Vanotti, M.B. Recovery of ammonia from poultry litter using gas permeable membranes. *Trans. ASABE* **2010**, *53*, 1267–1275. [CrossRef]
24. Majd, A.M.S.; Mukhtar, S.; Kunz, A. *Application of Diluted Sulfuric Acid for Manure Ammonia Extraction Using a Gas-Permeable Membrane*; American Society of Agricultural and Biological Engineers: St. Joseph, MI, USA, 2012. [CrossRef]
25. Majd, A.M.S.; Mukhtar, S. Ammonia Recovery Enhancement Using a Tubular Gas-Permeable Membrane System in Laboratory and Field-Scale Studies. *Am. Soc. Agric. Biol. Eng.* **2013**, *56*, 1951–1958. [CrossRef]
26. Blet, V.; Pons, M.N.; Greffe, J.L. Separation of ammonia with a gas-permeable tubular membrane. *Anal. Chim. Acta* **1989**, *219*, 309–311. [CrossRef]
27. Anthonisen, A.C.; Loehr, R.C.; Prakasam, T.B.S.; Srinath, E.G. Inhibition of nitrification by ammonia and nitrous acid. *J. Water Pollut. Control Fed.* **1976**, *48*, 835–852.
28. Emerson, K.; Russo, R.C.; Lund, R.E.; Thurston, R.V. Aqueous Ammonia Equilibrium Calculations: Effect of pH and Temperature. *J. Fish. Res. Board Can.* **1975**, *32*, 4. [CrossRef]
29. Lahav, O.; Mor, T.; Heber, A.J.; Molchanov, S.; Ramirez, J.C.; Li, C.; Broday, D.M. A new approach for minimizing ammonia emissions from poultry houses. *Water Air Soil Pollut.* **2008**, *191*, 183–197. [CrossRef]
30. Dube, P.J.; Vanotti, M.B.; Szogi, A.A.; Garcia-González, M.C. Enhancing recovery of ammonia from swine manure anaerobic digester effluent using gas-permeable membrane technology. *Waste Manag.* **2016**, *49*, 372–377. [CrossRef] [PubMed]
31. APHA; AWWA; WEF. *Standard Methods for the Examination of Water and Wastewater*, 21st ed.; American Public Health Association: Washington, DC, USA, 2005.
32. Venkataraman, K.; Choate, W.T.; Torre, E.R.; Husung, R.D.; Batchu, H.R. Characterization studies of ceramic membranes. A novel technique using a Coulter Porometer. *J. Membr. Sci.* **1988**, *39*, 259. [CrossRef]
33. SAS Institute. *SAS/STAT User's Guide, Ver. 9.2*; SAS Inst.: Cary, NC, USA, 2008.
34. Perfilov, V.; Fila, V.; Sanchez Marcano, J. A general predictive model for sweeping gas membrane distillation. *Desalination* **2018**, *443*, 285–306. [CrossRef]
35. Schneider, I.M.; Marison, W.; Stockar, U. Principles of an efficient new method for the removal of ammonia from animal cell cultures using hydrophobic membranes. *Enzym. Microb. Technol.* **1994**, *16*, 957–963. [CrossRef]
36. Daguerre-Martini, S.; Vanotti, M.B.; Rodriguez-Pastor, M.; Rosal, A.; Moral, R. Nitrogen recovery from wastewater using gas-permeable membranes: Impact of inorganic carbon content and natural organic matter. *Water Res.* **2018**, *137*, 201–210. [CrossRef]

37. Fillingham, M.; VanderZaag, A.C.; Singh, J.; Burtt, S.; Crolla, A.; Kinsley, C.; MacDonald, J.D. Characterizing the performance of gas-permeable membranes as an ammonia recovery strategy from anaerobically digested dairy manure. *Membranes* **2017**, *7*, 59. [CrossRef]

environments

MDPI

Article

Nitrogen Mineralization in a Sandy Soil Amended with Treated Low-Phosphorus Broiler Litter

Ariel A. Szogi *, Paul D. Shumaker, Kyoung S. Ro and Gilbert C. Sigua

U. S. Department of Agriculture-Agricultural Research Service, Coastal Plains Soil, Water and Plant Research Center, 2611 W. Lucas St., Florence, SC 20501, USA

* Correspondence: ariel.szogi@usda.gov; Tel.: +1-843-669-5203

Received: 15 July 2019; Accepted: 11 August 2019; Published: 14 August 2019

Abstract: Low-phosphorus (P) litter, a manure treatment byproduct, can be used as an organic soil amendment and nitrogen (N) source but its effect on N mineralization is unknown. A laboratory incubation study was conducted to compare the effect of adding untreated (fine or pelletized) broiler litter (FUL or PUL) versus extracted, low-P treated (fine or pelletized) broiler litter (FLP or PLP) on N dynamics in a sandy soil. All four litter materials were surface applied at 157 kg N ha^{-1}. The soil accumulation of ammonium (NH_4^+) and nitrate (NO_3^-) were used to estimate available mineralized N. The evolution of carbon dioxide (CO_2), ammonia (NH_3), and nitrous oxide (N_2O) was used to evaluate gaseous losses during soil incubation. Untreated litter materials provided high levels of mineralized N, 71% of the total N applied for FUL and 64% for PUL, while NH_3 losses were 24% to 35% and N_2O losses were 3.3% to 7.4% of the total applied N, respectively. Soil application of low-P treated litter provided lower levels of mineralized N, 42% for FLP and 29% for PLP of the total applied N with NH_3 losses of 5.7% for FLP for and 4.1% for PLP, and very low N_2O losses (0.5%). Differences in mineralized N between untreated and treated broiler litter materials were attributed to contrasting C:N ratios and acidity of the low-P litter byproducts. Soil application of treated low-P litter appears as an option for slow mineral N release and abatement of NH_3 and N_2O soil losses.

Keywords: organic nitrogen; mineralization; ammonia gas; nitrous oxide; nitrification; denitrification; manure; quick wash; poultry litter

1. Introduction

Most of the spent broiler litter is applied to soils as a source of plant nutrients for crop and forage production [1,2]. However, recurrent land application of broiler litter in regions with a high concentration of poultry farms is a major environmental concern because of nutrient buildup in soils to elevated levels. After soil application, a significant fraction of the organic N in broiler litter mineralizes into NH_4^+ and NO_3^-. Both inorganic N forms become available for plant use during the growing season but can be lost via leaching or surface runoff contaminating water resources [3]. In addition, a significant portion of surplus N from broiler litter is lost into the atmosphere through emissions of NH_3 and N_2O [4,5]. These environmental risks are leading to the development of technologies to manage nutrient-rich broiler litter that allow the recycling of nutrients as organic soil amendments or plant fertilizer materials.

Several management programs and technologies have been developed to solve the problem of surplus N and P from spent broiler litter including: (1) Transfer of broiler litter to nutrient-deficient agricultural lands as compost [6], as fine particles [7] or in pelletized form [8]; (2) improved manure application methods, such as subsurface soil placement of broiler litter, to prevent ammonia emissions or nutrient runoff [9,10]; (3) energy generation by thermal conversion such as incineration [11] or biological anaerobic digestion [12]; and (4) acidification with addition of chemicals to retain N in the broiler litter [13,14]. As an alternative, the U.S. Department of Agriculture developed a patented process, called "Quick Wash" (QW), to manage the surplus of N and P prior to soil application of broiler litter or animal manure [15]. The process uses a novel combination of acid, base, and organic polyelectrolyte to selectively extract a significant percentage of P from broiler litter while leaving most of the N in the organic, washed litter material. The QW approach has three distinctive advantages over other technologies or nutrient management strategies: (1) Compared with broiler litter transfer programs, there is no need to transport large volumes of broiler litter since only about 15% of its initial volume is shipped off the farm as a concentrated P product [16]; (2) compared to thermal conversion or the anaerobic digestion processes, the organic C and N in the treated low-P litter is conserved for soil health benefits; (3) compared to acidification processes such as alum addition, the treated low-P broiler litter can be safely land applied on a N basis because its N:P ratio is better balanced to match specific nutrient needs of crops.

Several studies have shown that the addition of broiler litter to soils can increase CO_2, NH_3, and N_2O gas emissions [17–19]. Broiler litter adds organic N along with organic C, stimulating mineralization of organic N and C with production of NH_4^+ and NO_3^- through microbial ammonification and nitrification, and N_2O through denitrification [20]. Therefore, slowing down nitrification and avoiding high NH_4^+ concentrations in the soil are important measures to lower N gaseous losses per unit of N input [5]. The objectives of the present study were to: (1) Compare if applications of low-P broiler litter treated with the QW process (hereafter called treated low-P litter) versus untreated broiler litter to a sandy soil would result in lower NH_3 and N_2O emissions; and (2) evaluate soil mineralization of low-P litter sources. To meet these objectives, we performed a laboratory soil incubation study in which both sources of N were surface applied to soil in un-pelletized and pelletized forms. The study included the determination of cumulative CO_2, NH_3, and N_2O emissions along with the soil concentrations of NH_4^+ and NO_3^- during the course of a laboratory incubation to evaluate the N mineralization of each N source. The study used a characteristic sandy soil common to areas with intense broiler production within the eastern Coastal Plains region, USA.

2. Materials and Methods

2.1. Soil Collection

Soil samples were collected from the topsoil (Ap horizon) of a Norfolk loamy sand (Fine-loamy kaolinitic thermic Typic Kandiudults) at the USDA-ARS Coastal Plains Soil, Water, and Plant Research Center in Florence, SC, USA. The area of the field used for the soil collection in this study was under conservation tillage with paratill subsoiling. To evaluate the distribution of nutrients and pH of the topsoil, composite soil samples were taken at 7.5 and 15.0 cm depth for routine soil testing according to Bauer et al. [21]. Thereafter, soil cores were collected from the topsoil using a soil core sampler (AMS, Inc., American Falls, ID, USA) equipped with a replaceable acrylic plastic cylindrical sleeve (5-cm diameter × 15-cm long).

2.2. Sources of Broiler Litter

The study included a total of four poultry litter materials (two untreated and two treated using the QW process): Fine-particle untreated litter (FUL); pelletized untreated litter (PUL); fine-particle low-P treated litter (FLP); and pelletized low-P treated litter (PLP). The PUL material was prepared by pelleting the FUL. The FUL material was collected from a farm with six 25,000-bird broiler houses in Lee Co., SC, USA. The broiler litter used for P extraction using the QW process was also collected from the same farm in a separate sampling campaign. Details about the collection and processing of broiler litter samples before and after QW treatment are further described by Szogi et al. [16]. Briefly, the QW process consists of three consecutive steps: (1) Wet P extraction; (2) P recovery; and (3) P recovery enhancement. The FLP is the solid product from Step 1 of the QW process in which P bound to poultry litter solids was extracted in solution using citric acid with a target pH of 4.5 at ambient temperature and pressure. The FLP solids were subsequently separated from the acid extract, dewatered, and air dried. Both PUL and PLP materials were obtained by pelleting FUL or FLP using a PP200 pellet mill equipped with a 6-mm die and roller set (Pellet Pros Inc., Davenport, Iowa, USA).

2.3. Incubations

Two separate but simultaneous laboratory incubation experiments were performed using two sets of 15 soil cores for each experiment. Each set of soil cores received the following treatments in triplicate: Un-amended control, FUL, PUL, FLP, and PLP. All broiler litter materials were applied to the soil on a total N basis of 89.6 mg N kg^{-1} soil which is equivalent to an application of 157 kg N ha^{-1} to non-irrigated, high yielding, corn application rates in South Carolina [22]. The equivalent oven dry mass of broiler litter applied to the soil cores to match the 89.6 mg N kg^{-1} were 0.0 g for the Control, 0.772 g for FUL, 0.776 g for PUL, 1.204 g for FLP, and 1.240 g for PLP. To optimize microbial activity, distilled water was added dropwise with a syringe and a needle to adjust soil moisture to 60% water filled pore space (WFP) after surface application of the litter treatments [23]. Both sets of cores were incubated for 10 weeks (68 days) at an average ambient temperature of 23 °C and 65% relative humidity.

One set of soil cores was used for sampling the soil weekly during the incubation. The soil cores were sampled to a depth of 12.7 cm using a 0.7-cm diameter rubber stopper borer as a sampling tool. Samples were freeze-dried prior to analysis to minimize N conversion and N gas losses during sample preparation for analysis [24]. The "soil sample" cores were covered with a black polyethylene sheet that allowed gaseous exchange but retarded water evaporation loss. The weight of the cores was inspected daily to make up for evaporation losses and maintain 60% WFP throughout the incubation.

The other set of soil cores was used to determine CO_2, N_2O and NH_3 gas emissions. Each soil core was enclosed in a 2.0-L PET (polyethylene terephthalate) plastic chamber with a threaded polyethylene lid. The lid had a port for periodic gas sampling and a 0.91-mm diameter vent to prevent pressure build-up above ambient atmospheric pressure inside the chamber. Five separate 5-mL gas samples were taken equally spaced across one hour (0, 15, 30, 45, and 60 minutes) to determine CO_2 and N_2O gas fluxes. The NH_3 gas was trapped as ammonium (NH_4^+) by passive diffusion [25] from the source (soil surface) into an 8-mL glass vial holding 5.0 mL of 0.2 M sulfuric acid. The acid trap was attached with a rubber band to the outside wall of the soil core with its open end at the same level as of the soil surface.

The time between two flux samplings varied throughout the experiment. Specific flux sampling times for CO_2 and N_2O were 1, 3, 5, 7, 9, 13, 16, 20, 27, 34, 41, 48, 55, 62, and 68 days from the initiation of the incubation. At each sampling time, the incubation chambers were uncapped and remained open for 2 h to allow headspace gas exchange with the ambient atmosphere, change-out of the NH_3 acid trap, and adjust the soil moisture to 60% WFP before recapping the chamber for another flux sampling. Gas sampling was done more frequently during the first three weeks (20 days) of incubation because the highest gas production was expected for all three measured gases (CO_2, N_2O, and NH_3). Thereafter, gas sampling was measured every seven days because the flux of N_2O and CO_2 was observed to slow down for all treatments.

2.4. Chemical Analysis

2.4.1. Soil and Broiler Litter Material Properties

With the exception of NH_4-N and NO_3-N, which were carried out in our laboratory, the soil chemical characterization was done at the Clemson University, Agricultural Service Laboratory, Clemson, SC, USA (Table 1). Total soil C and N were determined via thermal combustion, soil P and K were determined in Mehlich 1 extracts by inductively coupled plasma analysis, soil cation exchange capacity (CEC) was determined by the neutral ammonium acetate method, and soil pH was determined in 1:1 ratio soil/deionized water using a glass pH electrode [26].

Table 1. Chemical properties of the Norfolk loamy sand soil.

Depth cm	Cg kg^{-1}	Nmg kg^{-1}	NH_4-N mg kg^{-1}	NO_3-N mg kg^{-1}	P mg kg^{-1}	K mg kg^{-1}	CEC cmol kg^{-1}	pH
0–7.5	10.9	891	10.4	0.54	33	71	4.2	5.3
7.5–15	-	-	-	-	17	46	3.2	5.3

Soil and broiler litter samples were extracted with 2*M* KCl and analyzed for ammonium (NH_4-N) and nitrite plus nitrate (NO_2 + NO_3-N), hereafter called NO_3-N. Analysis of both NH_4-N and NO_3-N was carried out using an EL×800 microplate reader (Bio-Tek Instruments, Inc. Winooski, VT) set to 650 nm [27]. The total inorganic N (N_t) is defined as the sum of NH_4-N + NO_3-N.

All broiler litter materials were analyzed for both total Kjeldahl N (TKN) and total P (TP) after acid digestion using a Technicon auto-analyzer (Technicon Instruments Corp., Tarrytown, NJ, USA). Total C was quantified by combustion with an Elementar VarioMax CN analyzer (Elementar Americas Inc., Ronkonkoma, NY, USA). Broiler litter pH was measured in wet samples (1:1 solid to deionized water mixture) with a pH combination electrode.

2.4.2. Gas Analysis

Gas samples were injected into 10-mL headspace vials and analyzed for CO_2 and N_2O concentration on a Bruker Model 450-GC (Bruker Daltonics, Billerica, MA) gas chromatograph (GC) outfitted for greenhouse gas (GHG) analysis [28]. The GC was equipped with a model 1041 injector operated at 50 °C and 263 kPa which was connected to a 10-port gas sampling valve and pressure-actuated solenoid valve. Five mL of vial headspace was injected using a Combi-Pal auto-sampler equipped with a 5-mL headspace syringe. A portion of the sample was transferred onto a 1.8-m long by 1.6-cm outer diameter column packed with 80/100 mesh Hay Sep Q with a helium flow rate of 55 mL min^{-1}. The column was connected to a thermal conductivity detector (TCD) operated at 150 °C and with a filament temperature of 200 °C for CO_2 analysis. Another portion of the sample was split by means of the gas switching valve to another 10-port gas sampling solenoid valve and a portion of this sample was transferred to a 1.8-m long by 1.6-cm outer diameter silico-steel column also packed with 80/100 mesh Hay Sep Q with a N_2 carrier flow rate of 20-mL min^{-1} for N_2O analysis. This column was connected to an electron capture detector (ECD) operated at 300 °C. The GC oven was operated at 40 °C. Quantification of both CO_2 and N_2O was performed relative to an external standard curve for each gas. The NH_3-N captured in the acid trap samples were analyzed as NH_4-N by chemically suppressed cation chromatography [29].

2.5. *Data Analysis*

The gas flux from the soil cores was calculated by fitting the time series headspace gas concentrations with the quadratic regression model [30]. The magnitude of the flux was further corrected by determining the theoretical flux underestimation [31]. Cumulative CO_2 and N_2O emissions were estimated from the gas fluxes determined at the specific sampling time points throughout the incubation [32]. At the end of the 10-week study, the cumulative production of CO_2, N_2O-N, and NH_3-N were statistically analyzed using the ANOVA procedure of SAS version 9.4 (SAS Institute, Cary, NC, USA). Pairwise comparisons of treatment means were performed using the least square difference (LSD) option and were considered different when the probability values were $p < 0.05$.

A repeated measures analysis was conducted using the PROC MIXED procedure of SAS to evaluate how quickly the soil mineralization (NH_4-N and NO_3-N) responded to various broiler litter treatments during the incubation experiment [33]. A first-order autoregressive covariance structure in SAS was used to test the effects of Treatment (Trt), Time (Week), and Trt × Week interaction. The differences in response patterns were considered different among treatments when the probability of *F*-values was $p < 0.05$ for the interaction Trt × Week.

The nitrification rates for each poultry litter material were estimated using a model that describes the kinetics of transformation of NH_4-N into NO_3-N [34]. To quantify the accumulation of NO_3-N with time (t), the integrated form of the Verhulst equation was used:

$$NO_3\text{-}N = a/1 + (a/[NO_3\text{-}N]_0 - 1) \exp(-ak\,[t - t_0]) \quad (1)$$

where a is the asymptotic value of accumulated NO_3-N, k is a constant, $[NO_3\text{-}N]_0$ is the initial value of NO_3-N at time zero (t_0). The nonlinear procedure of Prism 7, GraphPad Software, Inc. (San Diego, CA, USA) was used to fit equation 1 to experimental soil NO_3-N data versus t. The maximal rate of nitrification was calculated as $Kmax = k \times a^2/4$.

The kinetics of N mineralization from application of poultry litter materials applied to soil were described by a first-order rate model [35]. The amount of N mineralized during the incubation study was evaluated using the equation:

$$N_t = N_O(1 - \exp[-kt]) + N_i \quad (2)$$

where N_t is the total inorganic N (NO_3-N + NH_4-N) concentration minus control concentrations, N_O is the potentially available organic N, k is a rate constant, t the is time of incubation, and Ni is initial N at $t = 0$. The nonlinear procedure of Prism 7, GraphPad Software, Inc. (San Diego, CA, USA) was used to fit Equation (2) to experimental soil $N_{t,\ t,}$ and N_O.

In addition, the available mineralized N as percent of total N added with each broiler litter treatment was estimated according to the following equation [36].

$$\text{Available mineral N (\%)} = [(N_t \text{ in treated soil} - N_t \text{ in control soil})/\text{total N added}] \times 100 \quad (3)$$

3. Results

3.1. Broiler Litter Materials Used in the Soil Incubation

The concentration and proportions of C, N, P and other constituents were different among the four broiler litter materials used in the study (Table 2). All chemical parameters of the untreated broiler litter, FUL or PUL, had values within the range of those reported by service laboratory analysis [37]. All four broiler litter materials had C contents (434–499 g kg^{-1}) within the expected range for poultry litter materials [38]. Total N concentrations were 55 to 66% higher in FUL or PUL versus FLP or PLP and also had higher NH_4-N and NO_3-N contents. The C:N ratio of FUL or PUL was almost half of FLP and PLP. Both FUL or PUL had a less balanced N:P ratio of <4.0 and basic pH. In contrast, the FLP and PLP had N:P ratios > 4.0 along with acidic pH, both resulting from the QW treatment [15].

Table 2. Chemical properties of the four broiler litter materials: Fine untreated litter (FUL), pelletized untreated litter (PUL), fine low-P treated litter (FLP), and pelletized low-P treated litter (PLP). Data are average of two samples on a dry weight basis.

Broiler Litter Material	C g kg^{-1}	N g kg^{-1}	P g kg^{-1}	NH_4-N mg kg^{-1}	NO_3-N mg kg^{-1}	C:N Ratio	N:P Ratio	pH
FUL	434	41.2	11.1	4891	1281	10.5	3.7	7.95
PUL	434	40.9	15.1	3618	1335	10.6	2.7	7.95
FLP	479	26.4	6.2	159	BD [1]	18.1	4.3	6.31
PLP	499	24.8	4.2	333	20	20.10	5.9	5.36

[1] Below detection.

3.2. Emissions of CO_2, NH_3, and N_2O during Soil Incubation

Analysis of variance indicated significant differences in cumulative CO_2, NH_3, and N_2O production among treatment combinations. Despite their contrasting chemical properties (Table 2) and particle size (pelletized versus unpelletized), the soil cumulative CO_2 production corrected by the CO_2 emissions of the control soil (g C kg^{-1} soil) was not significantly different ($p < 0.05$) for any of the four broiler litter treatments (Table 3). However, the higher percent of cumulative CO_2 emitted per unit of C added to soil with the FUL (61%) or PUL (56%) versus FLP (38%) or PLP (36%) can be attributed to all broiler litter materials being added to soil at a fixed N rate equivalent to 157 kg N ha^{-1}. Thus, on average, the C addition to soil with FLP or PLP was 58% of those applied using FUL or PUL.

Table 3. Cumulative CO_2, N_2O, and NH_3 emissions corrected by subtracting the emissions of the control during 10 weeks of incubation of the soil cores that received surface application of fine-particle untreated litter (FUL); pelletized untreated litter (PUL); fine-particle low-P treated litter (FLP); and pelletized low-P treated litter (PLP).

N Source	CO_2		NH_3		N_2O	
	g C kg^{-1}	% [1]	mg N kg^{-1}	% [1]	mg N kg^{-1}	% [1]
FUL	1.2a [2]	61a	21.5ab	24ab	3.0b	3.3b
PUL	1.1a	56a	31.0a	35a	6.8a	7.5a
FLP	1.5a	38a	5.1b	5.7b	1.7b	1.9b
PLP	2.1a	36a	3.7b	4.1b	0.4b	0.5b
$LSD_{0.05}$	1.2	34	25	28	3.1	3.5

[1] Percentage of emissions with respect to the total C or N applied per kilogram of soil. [2] Means followed by the same letter are not significantly different according to least square difference ($LSD_{0.05}$).

3.3. Nitrogen Mineralization

The evolution of NH_4-N and NO_3-N content during the 10-week "soil sample" incubation with surface applications of four broiler litter treatments and an unamended control are presented in Figure 1A,B. The rise in NH_4-N content started immediately after application of the broiler materials to soil (Figure 1A). On average, the highest soil NH_4-N contents occurred in the first week for FUL (53.4 mg kg^{-1}) and PUL (48.1 mg kg^{-1}) followed by FLP (37.0 mg kg^{-1}) while PLP (22.7 mg kg^{-1}) remained almost as low as the control (20.9 mg kg^{-1}). Thereafter, in Week 4 soil NH_4-N concentrations declined to levels similar to the control until the end of the 10-week incubation study. Simultaneously, NO_3-N concentrations started to rise in the third week of incubation for all treatments suggesting microbial nitrification of NH_4-N (Figure 1B).

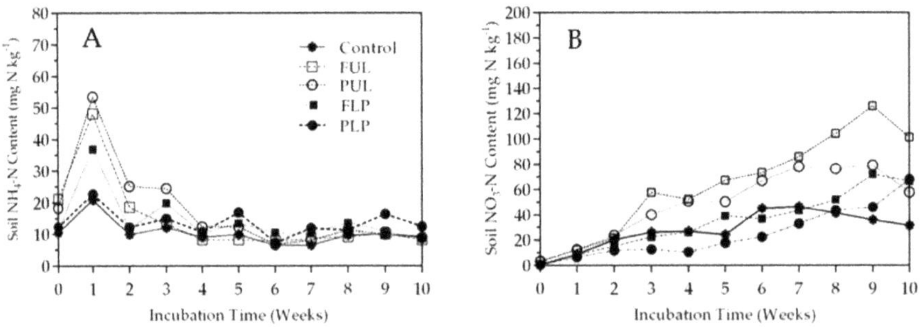

Figure 1. Evolution of the soil NH_4-N (**A**) and NO_3-N (**B**) content in the unamended control and surface applications to soil of fine-particle untreated litter (FUL); pelletized untreated litter (PUL); fine-particle low-P treated litter (FLP); and pelletized low-P treated litter (PLP). Each data point is the average of three replicates.

An ANOVA analysis indicated that there were significant differences among poultry treatments (Trt), time (Weeks), and the interaction Trt × Weeks for both soil NO_3-N and NH_4-N concentrations during the incubation study (Table 4). The analysis of Trt × Weeks effect confirmed that differences in

NH_4-N concentrations shown in Figure 1A among poultry litter treatment combinations during the first three weeks of incubation were statistically significant but differences were not significant with respect to the control on week 4 nor on any subsequent week until the end of the incubation study (combined Trt effect per week, Table 4). The combined Trt × Weeks effect analysis also confirms that NO_3-N concentrations in all treatments shown in Figure 1B were not significantly different to the soil control in the first two weeks of incubation (Table 4). Subsequently, NO_3-N concentrations started to rise in the third week of incubation showing significant differences among treatments until the end of the incubation study.

Table 4. Summary results of mineralization from repeated measures analysis: Analysis of variance and differences between the combined treatment (Trt) effect per week of soil incubation. Soil NH_4-N and NO_3-N, are presented with their corresponding average standard error of the mean (SEM) for the four treatment and control combination applied to Norfolk sandy loam soil and *p*-values.

Source		**NH_4-N**		**NO_3-N**	
	df	F-value	*p*-value [1]	F-value	*p*-value
Trt	4	8.7	0.0028 *	19.4	<0.0001 *
Week	10	83.4	<0.0001 *	22.1	<0.0001 *
Trt*Week	40	8.2	<0.0001 *	1.7	0.0309 *
		Combined Trt effect per week			
Time (Week)	df	SEM	*p*-value	SEM	*p*-value
0	4	14.4	0.0008 *	1.80	0.9995
1	4	36.5	<0.0001 *	9.1	0.9739
2	4	17.5	<0.0001 *	17.3	0.6836
3	4	16.9	0.0001 *	31.9	0.0056 *
4	4	10.3	0.7326	33.4	0.0052 *
5	4	12.1	0.0393	39.8	0.0011 *
6	4	7.8	0.7217	48.9	0.0005 *
7	4	9.0	0.4754	57.3	<0.0001 *
8	4	11.2	0.6946	63.5	<0.0001 *
9	4	11.5	0.1612	71.9	<0.0001 *
10	4	9.5	0.6677	65.2	<0.0001 *

[1] * Significant at the 0.05 probability level.

Soil NO_3-N accumulation as a result of biological nitrification was most rapid for the FUL and PUL treatments with maximal rates (*Kmax*) of 15.2 mg N kg^{-1} wk^{-1} and 14.3 mg N kg^{-1} wk^{-1}, respectively (Table 5). The faster soil NO_3-N accumulation for the FUL treatment could be attributed to its finer particle size and larger surface area than the pelletized litter material. Nitrification rates were much lower for FLP or PLP treatments with *Kmax* of 5.8 to 8.2 mg N kg^{-1} wk^{-1}. These lower *Kmax* suggested that properties of these materials besides particle size, such as low NH_4-N contents or C:N ratio, possibly had an effect on slower nitrification rates.

The addition to the soil of the four poultry litter materials initially increased the soil pH (Figure 2). At the onset of the incubation, the average soil pH was 5.03 but it rapidly increased in the first week to a value of 6.31 for the FUL, and in the second week to 6.78 for the PUL, possibly caused by the alkalinity of the materials (pH = 7.95, Table 2). In contrast, the soil pH increased only to values of 5.95 for the FLP and 5.88 for the PLP in the first week of incubation. These lower pH values were expected because of the acidic nature and low NH_4-N content of these materials (Table 2). After four weeks of incubation, soil pH declined to values below pH 5.0 in all treatments along with diminishing soil NH_4-N because of increasing microbial nitrification, an acid forming process. However, these lower pH values did not inhibit nitrification in the FUL and PUL treatments or fully explain the slow nitrification rates of the FLP and PLP treatments since the nitrification rates in acidic soils can equal or exceed those of neutral soils [39].

Table 5. Regression model parameters for the evolution of NO_3-N during the course of 10-week incubation of Norfolk soil amended with four broiler litter materials.

N Source	a [1]	K	$[NO_3]_0$	*Kmax*	R^2
			mg kg^{-1}	mg kg^{-1} wk^{-1}	
FUL	119	0.0043	11.3	15.2	0.93 [3]
PUL	73	0.0107	6.4	14.3	0.93 *
FLP	87	0.0043	7.2	8.2	0.95 *
PLP	– [2]	–	–	5.8	0.88 *
Control	–	–	–	3.5	0.64 *

[1] $NO_3\text{-N} = a/\{1 + (a/[NO_3\text{-N}]_0 - 1)\exp(-ak\,[t - t_0])\}$ where *a* is the asymptotic value of accumulated NO_3-N, k is a constant, $[NO_3\text{-N}]_0$ is the initial value of NO_3-N at time (t) zero, and *Kmax* is the maximal nitrification rate. [2] Data fitted a linear model: $NO_3\text{-N} = 5.84t - 4.39$ for PLP; and $NO_3\text{-N} = 3.50t + 10.4$ for Control where t is time in weeks.[3] Probability ($p = 0.05$) that the correlation of determination (R^2) is different from 0.

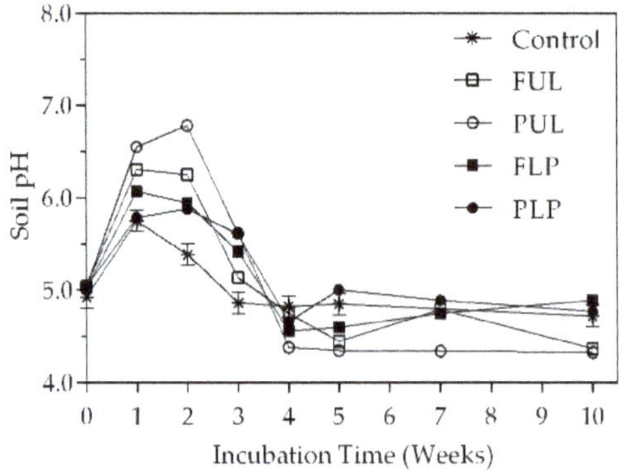

Figure 2. Evolution of soil pH during the soil incubation study.

The evolution of N_t during the 10-week incubation study and model equations for each broiler litter material applied to the Norfolk soil are presented in Figure 3. Except for PLP that had a linear response, the other three treatments had a typical non-linear response according to a first-order reaction model (Equation (2)). Both FUL and PUL had the greatest N_t production rates along with higher initial NH_4-N contents at time zero (Table 2) and rapid nitrification rates (*Kmax*) (Table 5). Instead, the N_t production rates were much lower for FLP or PLP most likely due to the very low levels of NH_4-N at time zero, slow mineralization and very slow nitrification rates (Figure 3). The available N_t during each week was used to estimate the inorganic N available as a percent of total N added with each broiler litter treatment (Table 6). The weekly inorganic N available estimates revealed the significant differences in available mineralized N between treatments in each of the first seven weeks of incubation. Thereafter, in the last three weeks of incubation, all four treatments had comparable percent of inorganic available N. For values above 100% available inorganic N, such as 124% for FUL in week 9, indicated a release of native soil N occurred during incubation [40]. On average, FUL (71%) and PUL (69%) had much higher mean percent available inorganic N than FLP (42%) and PLP (29%) suggesting FLP and PLP performed as a slow-release source of N_t.

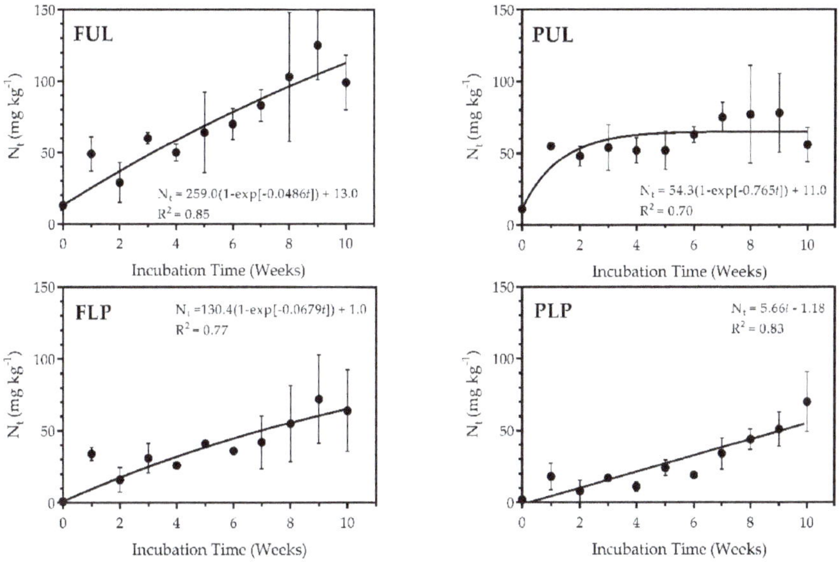

Figure 3. Total inorganic N (Nt = NO_3-N + NH_4-N) for four broiler litter materials applied to Norfolk soil: FUL = fine untreated litter; PUL = pelletized untreated litter; FLP = fine low-P treated litter; PLP = pelletized low-P treated litter. Except for PLP the lines are based on first order reaction (Equation). Data are the mean of three samples and error bars are one standard deviation of the mean.

Table 6. Weekly soil inorganic N available in percentage of total N added with four poultry litter materials applied on the surface of a Norfolk sandy loam soil. Data are the mean of three samples.

Time (Week)	FUL [1]	PUL	FLP	PLP
0	15a [2]	12a	1c	2d
1	55a	61a	38b	21c
2	32b	53a	18bc	9c
3	68a	60a	35b	19b
4	56a	58a	29b	12c
5	72a	58ab	46ab	27b
6	78a	70a	40b	21c
7	87a	84a	47b	38b
8	87a	86a	60a	49a
9	127a	87ab	80ab	57b
10	99a	78a	72a	63a
Mean	71	64	42	29
Standard deviation	31	21	23	20

[1] FUL = fine untreated litter; PUL = pelletized untreated litter; FLP = fine low-P treated litter; PLP = pelletized low-P treated litter. [2] For each week, the means with different letters are significantly different at the 0.05 probability level of the least square differences test (LSD).

4. Discussion

The Norfolk soil used for this study had the typical physio-chemical characteristics of a sandy soil under conservation tillage in the southeastern Coastal Plain region [21]. As a result of its conservation tillage management, the topsoil showed higher nutrient concentrations (total C, N, P, and K) in soil samples from 0 to 7.5 cm than from 7.5 to 15 cm depth (Table 1). Because of the low contents of NH_4-N and NO_3-N and uniform CEC and pH at both soil depths, we assumed that the 15-cm soil cores were adequate for evaluating the soil inorganic N dynamics of surface applications of raw or low-P broiler litter.

In general, at a soil moisture equivalent to 60% WFP, cumulative CO_2 emissions from microbial respiration increase to a maximum but denitrification and N_2O emissions should remain very low [23]. In our study, the cumulative CO_2 emissions had similar trends for all four litter treatments with no significant differences between pairwise mean treatment comparisons. As expected, the FUL, FLP, PLP treatment had the lowest cumulative N_2O emissions. However, the cumulative N_2O emission of the PUL treatment was significantly different and the highest for all treatments (3.0 mg N kg^{-1}). Since moisture of all soil cores was adjusted back to 60% WFP after each air headspace exchange, we observed that the fine-litter in the FUL and FLP were not wetter than the pellets in the PUL and PLP treatment since they crumbled in the first week of incubation as it would be the case of litter pellets applied to a crop field. Therefore, the significant differences in N_2O emissions could not be attributed either to differences in soil water content of the litter treatments, or particle size of the materials. Similar results to the FUL and PUL treatments were obtained by Cabrera et al. [41] on gaseous emissions from surface-applied fine-particle or pelletized broiler litter to sandy soils. At 58% WFP, their fine-particle broiler litter treatment showed maximum rates of CO_2 production but N_2O emissions rates were very low. In contrast, their pelletized broiler litter amendment produced similar CO_2 emissions as fine-poultry litter, but emitted significantly more N_2O than fine-particle litter. Likewise, Hayakawa et al. [42] reported that soil moisture at 50% WFP enhanced CO_2 emissions of soils amended with fine-particulate or pelletized poultry litter but promoted high N_2O production in the pelletized poultry litter treatment. They suggested that anaerobic conditions inside the pellets promoted denitrification during mineralization of the poultry litter pellets. Although FLP and PLP soil treatments had comparable cumulative CO_2 emissions to the FUL or PUL treatments, both had significant lower NH_3 and N_2O emissions likely because of key chemical properties such as C:N ratio and acidity.

Ammonia content at time zero and subsequent N mineralization of both FUL and PUL treatments released significant amounts of NH_4-N to soil but the alkaline pH in raw broiler litter raised the soil pH above 6.0 and induced significant N losses via NH_3 volatilization that ranged from 24 to 35% total N applied to soil (Table 3). Instead, the acidic pH of low-P broiler litter had a positive effect on abating soil NH_3 emissions from surface applied poultry litter by maintaining soil pH below 6.0 and decreasing NH_3 losses to less than 6% of the total N applied to soil. Acidifying amendments have been extensively studied for their effectiveness to decrease NH_3 emissions from broiler litter and their effect on microbial N mineralization [43]. Mineralization of organic N to NH_4-N requires enzymes produced by uric-acid degrading and urea-degrading soil microorganisms [44]. According to Burt et al. [45] acidification of broiler litter using flue-gas desulfurization gypsum impacted N mineralization resulting in lower pH, less NH_3 volatilization, and very low NO_3-N concentrations along with up to 57% decline in urea-degrading bacteria as compared to untreated broiler litter control. In our study, the acidic nature of FLP and PLP most likely slowed down the growth of urea-degrading bacteria or activities of urea-degrading enzymes which in turn slowed the mineralization of organic N to NH_4-N, and thereby, significantly abating NH_3 emissions with respect to FUL and PUL treatments.

Under the controlled conditions of soil moisture and temperature of our incubation study, the evolution of inorganic N content after addition of broiler litter materials to soil were also affected by the C:N ratio of each broiler litter product. In general, when organic materials with C:N ratios of less than 15:1 are added to soil, there is usually a rapid release of mineral NH_4^+ early in the mineralization process [46]. In our study, the rapid release of NH_4^+ and subsequent NO_3-N accumulation as a result of biological nitrification was most rapid for the FUL and PUL treatments with C:N ratio of 10.5:1. In contrast, C:N ratios higher than 25:1 promote N immobilization by soil microorganisms through the conversion of soil NH_4-N and NO_3-N into organic N lowering the inorganic N available for plant growth [20,46]. For C:N ratios in between 15 and 25, N mineralization is slow and immobilization may occur [47]. In our study the QW treated broiler litter materials, FLP and PLP, had C:N ratios > 15 that resulted in slow release of soil NH_4-N and NO_3-N with low NH_3 and N_2O emission losses.

The low-P treated broiler litter appears as a source capable of releasing N over an extended period and ideally could conserve soil N until it is needed by the crop while resolving the intractable problem of N and P imbalance in spent broiler litter. Being a slow available N source, low-P broiler litter may not effectively provide starter N during spring applications. However, it could be combined with other N sources such as untreated litter or synthetic fertilizers as starters to control N availability during the crop season, improving N use efficiency, and lessening the concerns of excessive N movement into water resources or the atmosphere. Longer incubation and field tests of using low-P broiler litter products as N sources appears warranted, especially field tests to evaluate the N use efficiencies during crop production.

Author Contributions: Conceptualization, A.A.S.; methodology, A.A.S., P.D.S., and K.S.R.; formal analysis, A.A.S., P.D.S., and G.C.S.; data curation, P.D.S.; writing—original draft preparation, A.A.S.; writing—review and editing, A.A.S., P.D.S., K.S.R. and G.C.S.

Funding: This research received no external funding.

Acknowledgments: This research was part of USDA-ARS National Program 212 Soil and Air, ARS Project 6082-12630-001-00D. Mention of trade names or commercial products in this article is solely for the purpose of providing specific information and does not imply recommendation or endorsement by the U.S. Department of Agriculture.

Conflicts of Interest: The authors declare no conflicts of interest.

References

1. Bolan, N.S.; Szogi, A.A.; Chuasavathi, T.; Seshadri, B.; Rothrock, M.J.; Panneerselvam, P. Uses and management of poultry litter. *World Poult. Sci. J.* **2010**, *66*, 673–698. [CrossRef]
2. Lin, Y.; Watts, D.B.; van Santen, E.; Cao, G. Influence of poultry litter on crop productivity under different field conditions: A meta-analysis. *Agron. J.* **2018**, *110*, 807–818. [CrossRef]
3. Bouwman, L.; Goldewijk, K.K.; Van Der Hoek, K.W.; Beusen, A.H.; Van Vuuren, D.P.; Willems, J.; Rufino, M.C.; Stehfest, E. Exploring global changes in nitrogen and phosphorus cycles in agriculture induced by livestock production over the 1900–2050 period. *Proc. Natl. Acad. Sci. USA* **2013**, *110*, 20882–20887. [CrossRef] [PubMed]
4. Powers, W.; Capelari, M. Measurement and mitigation of reactive nitrogen species from swine and poultry production. *J. Anim. Sci.* **2017**, *95*, 2236–2240. [PubMed]
5. Huang, T.; Gao, B.; Hu, X.-K.; Lu, X.; Well, R.; Christie, P.; Bakken, L.R.; Ju, X.-T. Ammonia-oxidation as an engine to generate nitrous oxide in an intensively managed calcareous Fluvo-aquic soil. *Sci. Rep.* **2014**, *4*, 3950. [CrossRef] [PubMed]
6. Sikora, L.J.; Enkiri, N.K. Comparison of phosphorus uptake from poultry litter compost with triple superphosphate in codorus soil. *Agron. J.* **2005**, *97*, 668–673. [CrossRef]
7. Paudel, K.P.; McIntosh, C.S. Country report: Broiler industry and broiler litter-related problems in the southeastern United States. *Waste Manag.* **2005**, *25*, 1083–1088. [CrossRef]
8. Bernhart, M.; Fasina, O.; Fulton, J.; Wood, C. Compaction of poultry litter. *Bioresour. Technol.* **2010**, *101*, 234–238. [CrossRef]
9. Pote, D.; Meisinger, J.J. Effect of poultry litter application method on ammonia volatilization from a conservation tillage system. *J. Soil Water Conserv.* **2014**, *69*, 17–25. [CrossRef]
10. Adeli, A.; McCarty, J.C.; Read, J.J.; Willers, J.L.; Feng, G.; Jenkins, J.N. Subsurface band placement of pelletized poultry litter in cotton. *Agron. J.* **2016**, *108*, 1356–1366. [CrossRef]
11. Bauer, P.J.; Szogi, A.A.; Shumaker, P.D. Fertilizer efficacy of poultry litter ash blended with lime or gypsum as fillers. *Environments* **2019**, *6*, 50. [CrossRef]
12. Marchioro, V.; Steinmetz, R.L.R.; do Amaral, A.C.; Gaspareto, T.C.; Treichel, H.; Kunz, A. Poultry itter solid state anaerobic digestion: Effect of digestate recirculation intervals and substrate/inoculum ratios on process efficiency. *Front. Sustain. Food Syst.* **2018**, *2*, 46. [CrossRef]
13. Fangueiro, D.; Hjorth, M.; Gioelli, F. Acidification of animal slurry—A review. *J. Environ. Manag.* **2015**, *149*, 46–56. [CrossRef] [PubMed]

14. Tomlinson, P.; Savin, M.; Moore, P., Jr. Long-term applications of untreated and alum-treated poultry litter drive soil nitrogen concentrations and associated microbial community dynamics. *Biol. Fertil. Soils* **2014**, *51*, 43–55. [CrossRef]
15. Szogi, A.A.; Vanotti, M.B.; Hunt, P.G. Process for Removing and Recovering Phosphorus from Animal Waste. U.S. Patent No. 8,673,046, 18 March 2014.
16. Szögi, A.A.; Vanotti, M.B.; Hunt, P.G. Phosphorus recovery from poultry litter. *Trans. ASABE* **2008**, *51*, 1727–1734. [CrossRef]
17. Cabrera, M.L.; Cheng, S.C.; Merka, W.C.; Pancorbo, O.C.; Thompson, S.A. Nitrous oxide and carbon dioxide emissions from pelletized in non-pelletized poultry litter incorporated into soil. *Plant Soil* **1994**, *163*, 189–196. [CrossRef]
18. Kim, S.U.; Ruangcharus, C.; Kumar, S.; Lee, H.H.; Park, H.J.; Jung, E.S.; Hong, C.O. Nitrous oxide emission from upland soil amended with different animal manures. *Appl. Biol. Chem.* **2019**, *62*, 8. [CrossRef]
19. Sharpe, R.; Schomberg, H.; Harper, L.; Endale, D.; Jenkins, M.; Franzluebbers, A. Ammonia volatilization from surface-applied poultry litter under conservation tillage management practices. *J. Environ. Qual.* **2004**, *33*, 1183–1188. [CrossRef]
20. Thangarajan, R.; Bolan, N.S.; Tian, G.; Naidu, R.; Kunhikrishnan, A. Role of organic amendment application on greenhouse gas emission from soil. *Sci. Total Environ.* **2013**, *465*, 72–96. [CrossRef]
21. Bauer, P.J.; Frederick, J.R.; Busscher, W.J.; Novak, J.M.; Fortnum, B.A. Soil sampling for fertilizer recommendations in conservation tillage with paratill subsoiling. *Crop Manag.* **2008**, *7*. [CrossRef]
22. Clemson University. *Regulatory Services, Soil Test Rating System*; Ag Service Lab, Clemson University: Clemson, SC, USA, 2019; Available online: https://www.clemson.edu/public/regulatory/ag-srvc-lab/soil-testing/pdf/rating-system.pdf (accessed on 13 July 2019).
23. Linn, D.M.; Doran, J.W. Effect of water-field pore space on carbon dioxide and nitrous oxide production in tilled and non-tilled soils. *Soil Sci. Soc. Am. J.* **1984**, *48*, 1267–1272. [CrossRef]
24. Mahimairaja, S.; Bolan, N.S.; Hedley, M.J.; McGregor, A.N. Evaluation of methods of measurement of nitrogen in poultry and animal manure. *Nutr. Cycl. Agroecosyst.* **1990**, *24*, 141–148. [CrossRef]
25. Lahav, O.; Mor, T.; Heber, A.J.; Molchanov, S.; Ramirez, J.C.; Li, C.; Broday, D.M. A new approach for minimizing ammonia emissions from poultry houses. *Water Air Soil Pollut.* **2008**, *191*, 183–197. [CrossRef]
26. Sikora, F.J.; Moore, K.P. *Soil Test Methods from the South Eastern United States*; SERA-IEG-6; Southern Cooperative Series Bulletin: Lexington, KY, USA, 2014.
27. Sims, G.K.; Ellsworth, T.R.; Mulvaney, R.L. Microscale determination of organic nitrogen in water and soil extracts. *Commun. Soil Sci. Plant Anal.* **1995**, *26*, 303–316. [CrossRef]
28. Duvekot, C. Analysis of greenhouse gases by gas chromatography. In *Agilent Technologies Application Note SI-01741*; Agilent Technologies: Santa Clara, CA, USA, 2010.
29. ASTM. Test method for determination of dissolved alkali and alkaline earth cations and ammonium in water and waste water by iron chromatography. In *Annual Book of Standards. VOL 11.02*; ASTM Standard D6919-09; American Society for Testing and Materials: Washington, DC, USA, 2009.
30. Wagner, S.W.; Reicosky, D.C.; Alessi, R.S. Regression models for calculating gas fluxes measured with a closed chamber. *Agron. J.* **1997**, *89*, 279–284. [CrossRef]
31. Venterea, R. Simplified method for quantifying theoretical underestimation of chamber—Based trace gas fluxes. *J. Environ. Qual.* **2010**, *39*, 126–135. [CrossRef]
32. Parkin, T.B.; Venterea, R.T. Sampling Protocols. Chamber-based trace gas flux measurements. In *Sampling Protocols*; Follet, R.F., Ed.; USDA-ARS Gracenet: Washington, DC, USA, 2010; Chapter 3; pp. 3-1–3-9.
33. Gezan, S.A.; Carvalho, M. Chapter 10: Analysis of repeated measures for the biological and agricultural sciences. In *Applied Statistics in Agricultural, Biological, and Environmental Sciences*; Glaz, B., Yeater, K.M., Eds.; American Society of Agronomy, Crop Science Society of America, and Soil Science Society of America, Inc.: Madison, WI, USA, 2018; pp. 279–297.
34. Hadas, A.; Feigembaum, S.; Feigin, A.; Portnoy, R. Nitrification rates in profiles of differently managed soil types. *Soil Sci. Soc. Am. J.* **1986**, *50*, 633–639. [CrossRef]
35. Havlin, J.L.; Beaton, J.D.; Tisdale, S.L.; Nelson, W.L. *Soil Fertility and Fertilizers. An Introduction to Nutrient Management*, 6th ed.; Prentice Hall: Upper Saddle River, NJ, USA, 1999.
36. Moore, A.D.; Mikkelsen, R.L.; Israel, D.W. Nitrogen mineralization of anaerobic swine lagoon sludge as influenced by seasonal tempertures. *Commun. Soil Sci. Plant Anal.* **2004**, *35*, 991–1005. [CrossRef]

37. Sharpley, A.; Slaton, N.; Tabler, T.; VanDevender, K.; Daniels, M.; Jones, F.; Daniels, T. *Nutrient Analysis of Poultry Litter*; University of Arkansas System, Agriculture and Natural Resources, Research and Extension: Little Rock, AR, USA, 2009.
38. Lynch, D.; Henihan, A.M.; Bowen, B.; Lynch, D.; McDonnell, K.; Kwapinski, W.; Leahy, J.J. Utilisation of poultry litter as an energy feedstock. *Biomass Bioenergy* **2013**, *49*, 197–204. [CrossRef]
39. Li, Y.; Chapman, S.J.; Nicol, G.W.; Yao, H. Nitrification and nitrifiers in acidic soils. *Soil Biol. Biochem.* **2018**, *116*, 290–301. [CrossRef]
40. Cayuela, M.L.; Velthoh, G.L.; Mondini, C.; Sinicco, T.; Van Groeningen, J.W. Nitrous oxide and carbon dioxide emissions during initial decomposition of animal by-products applied as fertilizers to soils. *Geoderma* **2010**, *157*, 235–242. [CrossRef]
41. Cabrera, M.L.; Chiang, S.C.; Merka, W.C.; Pancorbo, O.C.; Thompson, S.A. Pelletizing and soil water effects on gaseous emissions from surface-applied poultry litter. *Soil Sci. Soc. Am. J.* **1994**, *58*, 807–811. [CrossRef]
42. Hayakawa, A.; Akiyama, H.; Sudo, S.; Yagi, K. N_2O and NO emissions from an Andisol field as influenced by pelleted poultry manure. *Soil Biol. Biochem.* **2009**, *41*, 521–529. [CrossRef]
43. Cook, K.L.; Rothrock, M.J.; Eiteman, M.A.; Lovanh, N.; Sistani, K. Evaluation of nitrogen retention and microbial populations in poultry litter treated with chemical, biological or adsorbent aamendments. *J. Environ. Manag.* **2011**, *92*, 1760–1766. [CrossRef]
44. Rothrock, M.J.; Cook, K.L.; Warren, J.G.; Sistani, K. Microbial mineralization of organic nitrogen forms in poultry litter. *J. Environ. Qual.* **2010**, *39*, 1848–1857. [CrossRef]
45. Burt, C.D.; Cabrera, M.L.; Rothrock, M.J.; Kissel, D.E. Flue-gas desulfurization effects on urea-degrading bacteria and ammonia volatilization from broiler litter. *Poult. Sci.* **2017**, *6*, 2676–2683. [CrossRef]
46. Nahm, K.H. Factors influencing nitrogen mineralization during poultry litter composting and calculations for available nitrogen. *World Poult. Sci. J.* **2005**, *61*, 238–255. [CrossRef]
47. Nguyen, T.T.; Cavagnaro, T.R.; Ngo, H.T.T.; Marschner, P. Soil respiration, microbial biomass and nutrient availability in soil amended with high and low C/N residue–Influence of interval between residue additions. *Soil Biol. Biochem.* **2016**, *95*, 189–197. [CrossRef]

environments

MDPI

Article

Poultry Litter, Biochar, and Fertilizer Effect on Corn Yield, Nutrient Uptake, N_2O and CO_2 Emissions

Karamat R. Sistani [1,*], Jason R. Simmons [1], Marcia Jn-Baptiste [2] and Jeff M. Novak [3]

1 USDA-ARS, 2413 Nashville Rd B-5, Bowling Green, KY 42101, USA; jason.simmons@usda.gov
2 Self-Employed, Bowling Green, KY 42104, USA; marcia.jeanbaptiste@gmail.com
3 USDA-ARS, Florence, SC 29501, USA; jeff.novak@usda.gov
* Correspondence: karamat.sistani@usda.gov; Tel.: +1-270-779-4004

Received: 21 April 2019; Accepted: 23 May 2019; Published: 24 May 2019

Abstract: Biochar holds promise as a soil amendment with potential to sequester carbon, improve soil fertility, adsorb organic pollutants, stimulate soil microbial activities, and improve crop yield. We used a hardwood biochar to assess its impact on corn (*Zea mays*) grain, biomass yields and greenhouse gas emission in central Kentucky, USA. Six treatments included as follows: control (C) with no amendment applied; poultry litter (PL); biochar (B); biochar + poultry litter (B + PL); fertilizers N-P-K (F); and biochar + fertilizers (B + F). Biochar was applied only once to plots in 2010 followed by rototilling all plots. Only PL and fertilizer were applied annually. When applied alone, biochar did not significantly increase dry matter, grain yield, and N-P-K uptake. There was also no significant difference between the combined treatments when compared with PL or F applications alone. We observed a slight increasing trend in corn grain yield in the following 2 years compared to the first year from biochar treatment. Poultry litter treatment produced significantly greater N_2O and CO_2 emissions, but emissions were lower from the B+PL treatment. We conclude that this biochar did not improve corn productivity in the short term but has potential to increase yield in the long term and may have some benefit when combined with PL or F in reducing N_2O and CO_2 emissions.

Keywords: animal manure; poultry litter; biochar; corn; greenhouse gas; nutrient

1. Introduction

One strategy that has been advocated for mitigating and reducing global CO_2 concentration is based on the pyrolysis of biomass—a process that produces a byproduct known as biochar. When biochar is produced from biomass, it represents a net withdrawal of CO_2 from the atmosphere [1]. Biochar is black carbon, but not all black carbon materials are biochar [2]. The C in biochar is highly resistant to microbial degradation for many years [3]. It has also been emphasized that biochar holds great promise as a soil amendment to sequester carbon, improve soil fertility, adsorb organic pollutants, and stimulate soil microbial activities. Additionally, there are other benefits of incorporating biochar into a soil such as increases in cation exchange capacity (CEC) [4], increases in nutrient retention and availability for plant uptake particularly in highly weathered soils (Ultisols), and increases in fertilizer N use efficiency [5–8]. Clough and Condron [9] reported that agronomic benefits of biochar addition are well documented for highly weathered soils; however, little information exists on biochar impact on soil properties and crop yields in moderately weathered soils (Alfisols).

Biochar adds C structures into the stable soil organic matter (SOM) pool which improves soil fertility and crop productivity. The addition of biochar to soils may produce immediate effects on soil properties such as soil nutrition, water retention, and microbial activities [10–12]. Although these effects vary depending on soil type, the impacts may be long-term on soil and environments [13,14]. The addition of biochar to soils has not proven to be consistent with increasing crop yields. Approximately 50% of the compiled studies observed short-term positive yield or growth impact, while 30% have reported no significant differences and 20% reported negative yield or growth effects [2]. They also reported that greater positive yield impacts for biochar addition occurred when it was applied to highly weathered or degraded soils with limited fertility and productivity. Novak et al. [15,16] reported that biochar produced at higher pyrolysis temperatures increased soil pH. They also concluded that biochar produced from different feedstocks under different pyrolysis conditions affects soil physical and chemical properties in different ways. Jeffery et al. [17] and Crane-Droesch et al. [18] evaluated the impact of various biochar on crop yield. They used results from two meta-analyses of biochar studies and showed an overall average crop yield increase of 10% after applying biochar to soil. They also reported that crop yields were soil-type-dependent and growth enhancements were highly variable. Another more recent meta-analysis using data from 84 reports by Crane-Droesch et al. [18] reported that biochar increased crop yields in the highly weathered soils of humid tropics more than in nutrient-rich temperate soils. One reason for crop yield variability in these studies may be due to differences in biochar quality associated with the feedstock, pyrolysis conditions, application rates, and soil properties [19]. For example, yield improvements were often associated with biochar application to low pH, nutrient-poor soils because alkaline biochar induced a soil-liming effect [20,21]. The array of controversies regarding the benefits of biochar and the variabilities reported for similar treatments through different studies as reported in the literature warrants further and continued research with different types of biochar for specific regions and crops. Therefore, the objective of this study was to investigate the impact of hardwood biochar alone and in combination with poultry litter or with chemical fertilizer on corn growth, grain yield, and greenhouse gas emissions on a Crider silt loam near Bowling Green, Kentucky USA. We hypothesized that the hardwood biochar used in this study, whether applied alone or combined with PL, would increase corn productivity and reduce N_2O and CO_2 emissions.

2. Material and Methods

Field experiments were conducted in 2010, 2011, and 2013 (the experiment was lost in 2012 due to severe summer drought) in which no-till corn was grown for grain on a Crider silt loam soil (Fine-silty, mixed, active, mesic, Typic Paleudalfs) of 1–2% slope in Bowling Green, Kentucky, Kentucky, USA (36°56′19.1″ N, 86°28′51.2″ W), with textural analysis of 3.1% sand, 65.3% silt, and 31.6% clay and soil organic matter of 25 g kg^{-1}. The region has a temperate climate with a typical mean temperature of 14.5 °C and rainfall of 1300 mm $year^{-1}$. Precipitation data were collected from a nearby weather station (Western Kentucky University Research Farm) during the growing seasons. The experimental design was a randomized complete block with three replicated plots (4.6-m × 3.5-m). The study site was fallow under mixed grasses such as tall fescue, orchardgrass, and white clover. We purposely chose the site because it had not been fertilized for the previous 5 years. The study included six treatment controls (C) with no amendment applied, poultry litter (PL) (a mixture of poultry manure and bedding materials) applied at a rate to provide 224 kg N ha^{-1}; biochar (B) applied alone at the rate of 21.28 Mg ha^{-1}; biochar + poultry (B + PL); chemical fertilizers; N-P-K (F); and biochar + chemical fertilizers (B + F). Every PL (alone or combined) was surface broadcast applied at the rate to provide 224 kg N ha^{-1} assuming 50% of the PL total N becomes plant-available during the growing season. Poultry litter application also supplied 245, 181, 187, and 376 kg P ha^{-1}, and 599, 447, 484, and 747 kg K ha^{-1} in 2010, 2011, 2012, and 2013, respectively. Due to the unbalanced nutrient content of the poultry litter, particularly N:P:K compared to the N:P:K requirement of corn, there was a surplus of P and K applied from litter treatment, since the poultry litter quantity was applied base on the N requirement of the corn. The chemical fertilizers consisted of a blend of urea, muriate of potash, and diammonium phosphate (46-0-0, 0-0-60, 18-46-0). The fertilizer blend was surface broadcast applied at the rate of 224 kg N ha^{-1} plus P and K based on soil test recommendations, 67.2 kg P ha^{-1} and 112 kg K ha^{-1} respectively.

Soil samples (background) were taken to a depth of 15 cm before initial treatment applications. Five random cores of soil were taken with a 2.54-cm diameter soil probe from each plot, combined in a plastic zip seal bag, air-dried, and ground with a Dynacrush soil crusher (Custom Laboratory Inc.,

Holden, MO, USA) to pass a 2.0-mm mesh. The soil chemical properties were as follows: pH (4.7) measured on a 1:1 soil:$CaCl_2$ (0.05 *M*) solution using a combination electrode (Accuphast electrode, Fisher Scientific, Pittsburg, PA, USA), total N (1.4 g kg^{-1}) and TC (13.05 g kg^{-1}) in soil were measured using a Vario Max CN analyzer (Elementar Americas, Inc. Mt. Laurel, NJ, USA), and NH_4-N (22.8 mg kg^{-1}) and NO_3-N (0.84 mg kg^{-1}) were extracted from soil samples with 2 *M* KCl [22] and then analyzed using flow injection analysis (QuickChem FIA+, Lachat Instruments, Milwaukee, WI, USA). Soil nutrient availability was also assessed after extraction with Mehlich 3 [23], and then elements were quantified using Inductively Coupled Plasma–Optical Emissions Spectroscopy (ICP-OES; Varian, Vista Pro; Varian Analytical Instruments, Walnut Creek, CA, USA). The elements analyzed were as follows: P 4.88 mg kg^{-1}; Ca 1334 mg kg^{-1}; Mg 117 mg kg^{-1}; K 152 mg kg^{-1}; Na 79 mg kg^{-1}; Fe 87 mg kg^{-1}; Mn 144 mg kg^{-1}; and Zn 1.85 mg kg^{-1}.

Approximately 340 kg of CQuest™ biochar for this experiment was produced using sawdust generated from the wood flooring process using mixed hardwood species subject to a fast pyrolysis process at 500–600 °C by the Dynamotive Technologies Corp., (West Lorne, ON, Canada). The biochar was pulverized to fine-sized (<0.5-mm) material and stored in steel drums. Chemical and physical analysis of the hardwood biochar is presented in Table 1. The % moisture, ash content, fixed C, volatile C, and elemental (C, H, O, and N) contents were determined on an oven dry-weight basis by Hazen Research, Inc. (Golden, CO, USA) following ASTM D 3171 and 3176 standard methods (ASTM, 2006). In this method, the O content was determined by difference. These results were used to calculate a molar O/C and H/C ratio. The total K, Ca and P contents were determined using the USEPA 3052 microwave-assisted acid digestion method (USEPA, 1996) and their concentrations were quantified using an inductively-coupled plasma mass spectrometer as outlined by Ref. [24]. The hardwood biochar pH, specific surface area (SAA), and total acidity were measured as outlined by Ref. [15]. The hardwood biochar was surface broadcast applied to all B treatments and incorporated to soil to a depth of 10 cm by a rototiller to prevent potential surface transport of biochar out of the plots by wind or water. The biochar was applied only once at the start of this study in the spring of 2010.

Table 1. Characteristics of hardwood biochar (oven dry basis, SD = standard deviation).

Property	Mean	SD
%H_2O	4.6	0.61
%ash	14.16	13.21
%Fixed C	46	8.49
%Volatile C	54	26
%C	68.29	6.86
%H	2.67	0.65
%N	0.25	0.09
%O*	13.38	3.86
O/C	0.16	0.02
H/C	0.51	0.04
%Ca	0.49	0.06
%K	0.65	0.05
%P	0.03	na
pH (H_2O)	5.59	0.61
SSA (m^2/g) †	1.29	na
Total acidity (cmol/100g)	120	na

† where SAA = specific surface area, and na = not available.

Corn (*Zea mays*) (DeKalb DKC61-69 RR/BT) was planted (76 cm row spacing) in 4.6 m × 3.5 m sized plots (total of six rows per plot) on 12 May 2010; 9 May 2011, and 2 May 2013. Total aboveground plant biomass was determined at physiological maturity (R6 growth stage) on 17 August 2010, 4 August 2011, and 14 August 2013. For biomass measurement, six plants were randomly selected per plot from a non-border row, cut above the soil surface, weighed, and shredded with a wood chipper (Modern

Tool and Die Company, Cleveland, OH, USA) to allow homogenous mixed subsamples to be taken. Samples were then oven-dried at 65 °C for a minimum of 72 h and ground into composite samples with a grinder mill (Thomas Wiley Mill Model 4; Thomas Scientific, Swedesboro, NJ, USA) to pass a 1-mm mesh. Corn was harvested as grain on 9 September 2010, 23 August 2011, and 25 September 2013, by hand-picking the two center rows of each plot for a harvest area of 1.5 m × 3.5 m. Grain subsamples from each plot were taken for analyzing moisture content. Samples were oven-dried at 65 °C and ground with a grinder mill to pass a 1-mm mesh for chemical analysis. Plant tissue samples were analyzed for total N and C using a CN analyzer, and a dry ash/acid extraction procedure [25] was used to analyze samples for total P and K using ICP-OES.

Gaseous emissions (N_2O and CO_2) were measured during the growing seasons using static, vented chambers [26] and a gas chromatograph analyzer. Measurements of the N_2O and CO_2 fluxes were taken from May (planting) to September (harvest) for only 2 years (2010 and 2011) following the procedures reported by Mosier et al. [27]. Measurements were generally taken two to three times per week midmorning of each sampling day. The chambers used were made of aluminum and measured 10 cm tall. After treatment applications, anchors were forced into the ground to a depth of 15 cm in each plot such that they were flush with the soil surface. Anchors were installed each year, 1 to 3 days prior to beginning measurements and were not removed until the fall after harvest. At each flux measurement time, the chambers were placed on fixed anchors (38 cm wide and 102 cm long). The anchors were placed such that the 102 cm length was parallel to the corn rows. Plants emerging inside the measurement area were removed. Flux measurements sites were included within each replicate of each treatment plot. Air samples (40 mL) were collected from inside the chambers by syringe at 0, 15, and 30 min intervals after the chambers were seated on the anchors. The air samples were injected into 20 ml evacuated vials that were sealed with grey butyl rubber septa. Samples were analyzed with a gas chromatograph (CP-3800, Varian, Inc., Palo Alto, CA, USA) equipped with a thermoconductivity detector and an electron capture detector for quantification of CO_2 and N_2O. A quality control standard sample (known concentration) was also analyzed after every 25 unknown samples' analysis. Fluxes were calculated for each gas from the linear or non-linear [28] increase in concentration (selected according to the emission pattern) in the chamber headspace with time as suggested by Livingston and Hutchinson [26]. To calculate the cumulative growing season fluxes, estimates of daily N_2O and CO_2 emissions between sampling days were calculated using a linear interpolation between adjacent sampling dates.

The experimental design was a randomized complete block with three replicated plots (4.6-m × 3.5-m). Differences among treatments and years were determined by analysis of variance (ANOVA) using PROC GLM procedure (SAS Institute, 2001). Blocks were considered as random factor and year as repeated measurement. All statistical comparisons were made at $\alpha = 0.05$ probability level using Fisher's protected LSD to separate treatment means.

3. Results and Discussion

3.1. Environmental Condition

The 3-year experiment was established in 2010 with a corn planting date of 20 May. In 2012, the experiment was completely lost due to severe drought particularly during the critical growing season months of June, July, and August; hence, we continued the study for 1 more year in 2013. The total precipitation for the three critical months of June, July, and August were 309 mm, 286 mm, and 473 mm for 2010, 2011, and 2013, respectively (Figure 1). However, the distribution of the rainfall among the critical growing season months was better in 2013 followed by 2011 and 2010.

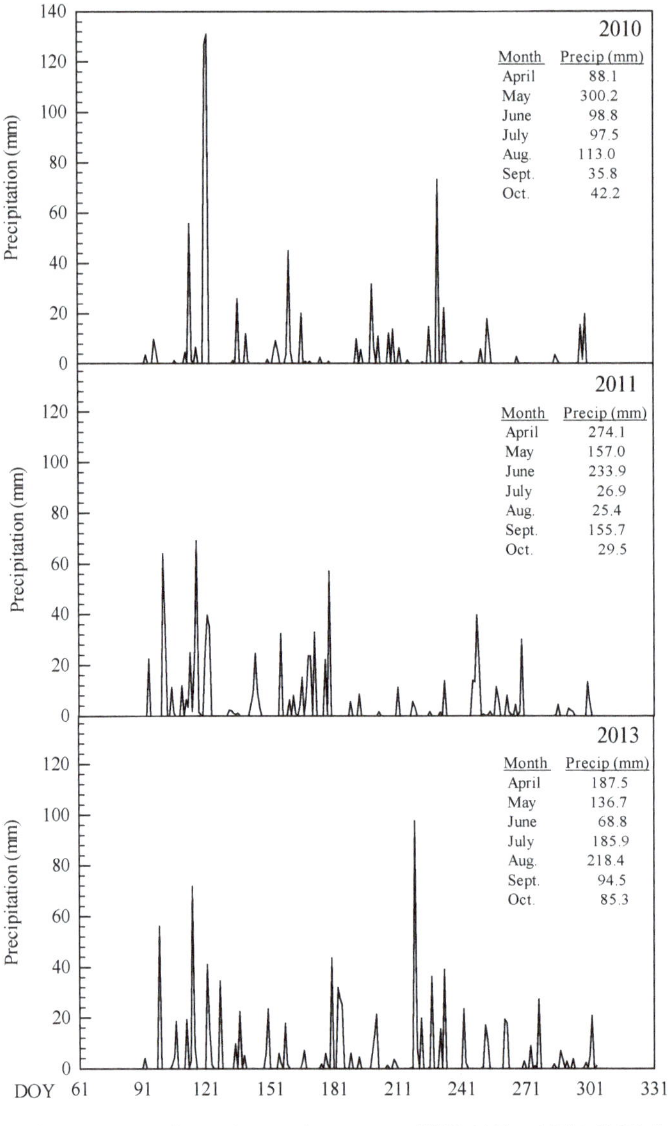

Figure 1. Daily precipitation during the growing seasons of 2010, 2011 and 2013. DOY, day of year.

3.2. Corn Dry Matter Yield

Since the interaction between year and all the agronomic measurements (dry matter yield, grain yield, and nutrient uptake) were significant, the results are reported on a yearly basis. No significant corn dry matter yield differences occurred among treatments in 2010 (dry year). In 2011, PL treatment produced the greatest dry matter yield and was significantly different than control, biochar, and fertilizer (N-P-K) treatments. The combination of PL and biochar resulted in significantly higher dry matter yield compared to biochar alone in the 2011 growing season (Table 2, Figure 2). In 2013, fertilizer, PL, and their combinations with biochar treatments produced significantly greater dry matter yields than control and biochar alone treatments. Similar to 2011, the addition of nutrients by way of fertilizer or PL enhanced the impact of biochar on corn dry matter production particularly under more favorable soil moisture conditions. We believe the level of impact of these treatments on corn dry matter yield was mostly related to the precipitation and soil moisture availability in each year. Results suggest that biochar alone did not contain the quantity and specific forms of required plant nutrients for optimum corn growth in this study. However, above ground dry matter yield increased when biochar was supplemented with nutrients from poultry litter and fertilizer under optimum soil moisture content compared to biochar alone treatment (Table 2, Figure 2). This is indicative of the synergistic effect of biochar when applied with fertilizer sources. Such a synergistic effect has been reported in various studies such as for the increased production of maize by Yamato et al. [29] through combined biochar and fertilizer; Yamato et al. reported a 4–12 times higher yield of rice and sorghum when combined with fertilizer and compost [14,30] and reported increases in corn yield through increased P availability and uptake when biochar combined with arbuscular mycorrhiza (AM) fungal spores [31,32].

Table 2. Whole plant dry matter and grain yield influenced by different treatment and growing season year.

	Whole Plant Dry Matter Yield			**Grain Yield**		
Treatment (T)	**2010**	**2011**	**2013**	**2010**	**2011**	**2013**
			kg ha^{-1}			
Control	16,262a †	18,100bc	13,995b	6480a	7201a	8952bc
Biochar (B)	13,828a	17,168c	14,090b	5619ab	6629a	8042c
Fertilizer (F)	17,379a	18,291bc	24,721a	5382ab	6036a	12,632a
Poultry Litter (PL)	16,581a	21,495a	24,245a	5037b	7062a	13,163a
B + F	13,875a	19,145abc	22,526a	4706b	5430a	11,723ab
B + PL	15,874a	20,524ab	22,170a	4721b	6281a	13,114a
$LSD_{(0.05)}$	4454	2920	6276	1232	1825	3557

† Values within columns followed by the same letters are not significantly different according to Fisher's LSD (0.05) level.

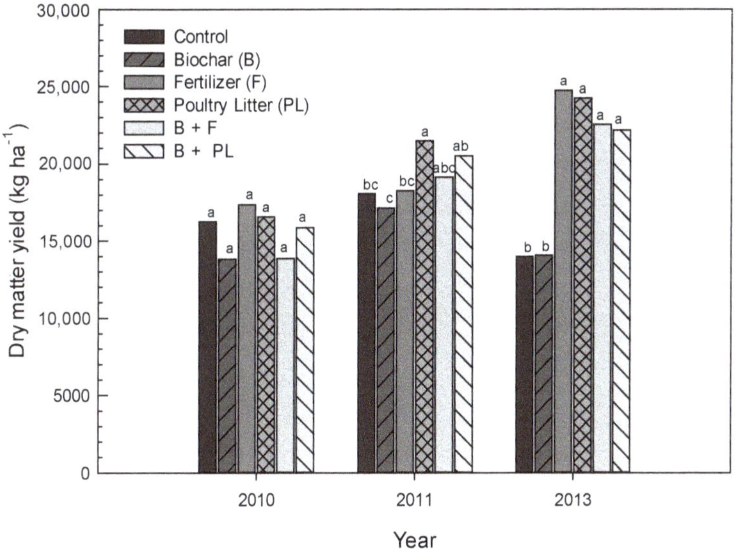

Figure 2. Effect of biochar, poultry litter, chemical fertilizer, and the combined treatments tested on corn dry matter yield in 2010, 2011, and 2013 growing seasons. Within each year, values followed by the same letters are not significantly different according to Fisher's LSD (0.05) level.

3.3. Corn Grain Yield

Significant differences in corn grain yield occurred among treatments in 2010 and 2013 growing seasons with 2013 producing the greatest, followed by 2011 and 2010 (Table 2, Figure 3). All treatments produced a similar grain yield in 2011 while biochar alone and control treatments had the lowest grain yield in 2013. When applied alone, biochar treatment did not increase corn grain yield in the three growing seasons compared to other treatments; however, the results indicate that over time, biochar positively impacted grain yield (Table 2), with a significant increase each consecutive year compared to the first year, which seems to indicate that biochar has potential to positively impact yield in the long term. This result is indicative of the carrying over effect of biochar also reported in a study by Adejumo et al. [33]. We also believe that the low soil pH may have influenced nutrient availability and uptake by corn. We also speculate that the low soil pH may have created a situation for Al toxicity for corn. Besides, organic matter is known to release nutrients slowly over time with only a small fraction of nutrients available the first year of application [34,35]. Biochar is also known to have a long-term effect on nutrient availability which is due to its potential to increase cation exchange capacity [6], consequently leading to increased yield over time. The control treatment was the same or higher than other treatments under limited moisture availability in 2010 and 2011. It seems to indicate that the addition of more nutrients through fertilizer and PL application exerted more stress and had a slightly negative impact on corn plants under drought conditions of 2010 and 2011 growing seasons (Table 2, Figure 3).

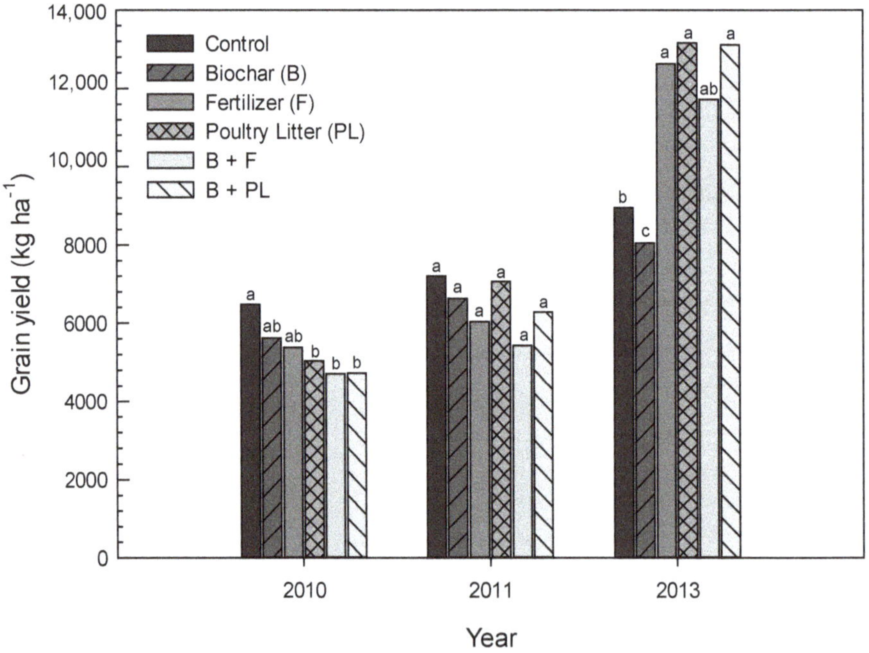

Figure 3. Effect of biochar, poultry litter, chemical fertilizer, and the combined treatments tested on corn dry matter yield in 2010, 2011, and 2013 growing seasons. Within each year, values followed by the same letters are not significantly different according to Fisher's LSD (0.05) level.

3.4. Corn Biomass N, P, and K Concentration and Uptake

The whole plant biomass N, P, and K concentrations were measured yearly at plant maturity followed by total N, P, and K uptake determination for each treatment. Since year by treatment interactions were mostly significant for N, P, and K biomass concentration and N, P, and K uptake, the results are shown separately by year in Tables 3 and 4. The biomass N concentration was significantly different among treatments in each year (Table 3). In more favorable soil moisture conditions (2011 and 2013), biochar treatment (alone) had lower biomass N concentration than other treatments except for control treatment; however, in 2010, a dry year, biochar alone had higher N concentration than when biochar alone was applied in the other years. In all three growing seasons, the biomass N content increased significantly when biochar was combined with nutrients by way of fertilizer or poultry litter compared to the biochar (alone) application (Table 3). The 2013 biomass N concentration was lower for all treatments compared to 2011 and 2010, probably due to greater biomass yield because of optimum soil moisture condition and better N use efficiency in 2013. The biomass P and K concentrations were significantly different among treatments except for PL and B + PL treatments, which had numerically higher P and K concentrations (Table 3). Table 4 shows N, P, and K uptake by corn for each growing season. No significant differences regarding N uptake among treatments were noted in the 2010 growing season. However, all treatments had significantly greater N uptake than biochar alone and control treatments in 2011 and 2013 growing seasons. In general, no specific trend regarding P and K uptake occurred in any of the growing seasons except for the B+PL treatment which had significantly greater P and K uptake than biochar alone treatment, indicating the low level of available nutrients of the biochar (Table 4). We speculate that precipitation and soil moisture availability may have had a big impact on corn nutrient uptake. Biochar in combination with fertilizer or poultry litter treatments positively influenced the N, P, K uptake more than biochar treatment alone, indicating a synergistic effect of the combined treatments [14,29,30]. Corn plants which were grown on poultry litter alone as a source of nutrients utilized N, P, and K more efficiently than other treatments (Table 4). It should be noted that even though the amount of P and K applications for the fertilizer treatment were based on the standard soil test recommendations, the poultry litter treatment provided more P and K because the quantity of poultry litter application was calculated based on the N requirement of the corn.

Table 3. Whole plants nutrients (N, P, and K) concentrations influenced by different treatments and growing season year.

	Whole Plant N, P, and K Concentration								
	N			P			K		
Treatment (T)	**2010**	**2011**	**2013**	**2010**	**2011**	**2013**	**2010**	**2011**	**2013**
					$g\ kg^{-1}$				
Control	11.1c †	7.9d	6.8b	1.3c	1.0d	1.3c	8.9b	13.8b	8.2b
Biochar (B)	12.9b	8.5d	7.3b	1.5bc	1.2cd	1.4c	10.6b	14.1b	8.7b
Fertilizer (F)	13.2ab	12.5ab	9.6a	1.5bc	1.4bc	1.6b	9.7b	15.5ab	8.6b
Poultry Litter (PL)	13.9ab	11.6bc	9.9a	2.0a	1.8ab	2.0a	11.8ab	18.5a	12.7a
B + F	14.0a	13.3a	9.2a	1.7b	1.5b	1.6b	11.4ab	17.6a	8.7b
B + PL	14.2a	10.7c	9.7a	2.1a	1.9a	2.0a	13.8a	18.9a	13.3a
$LSD_{(0.05)}$	1.09	1.35	0.99	0.26	0.35	0.21	2.99	3.40	3.29

† Values within columns followed by the same letters are not significantly different according to Fisher's LSD (0.05) level.

Table 4. Whole plants nutrients (N, P, and K) uptake influenced by different treatment and growing season year.

	Whole Plant N, P, and K Uptake								
	N			P			K		
Treatment (T)	**2010**	**2011**	**2013**	**2010**	**2011**	**2013**	**2010**	**2011**	**2013**
					kg ha^{-1}				
Control	182a †	144b	95b	21.0b	16.5d	17.6b	150b	250c	116c
Biochar (B)	178a	146b	103b	20.6b	20.2cd	19.1b	145b	243c	123c
Fertilizer (F)	230a	228a	239a	26.5ab	25.8bc	40.1a	170ab	283bc	212b
Poultry Litter (PL)	231a	249a	240a	33.9a	37.8a	48.5a	195ab	398a	306a
B + F	194a	254a	207a	23.0b	29.5b	36.0a	159ab	337ab	195b
B + PL	225a	220a	215a	32.5a	39.0a	44.8a	217a	390a	287a
$LSD_{(0.05)}$	62.6	37.5	68.0	9.40	6.86	14.08	66.7	75.0	53.9

† Values within columns followed by the same letters are not significantly different according to Fisher's LSD (0.05) level.

3.5. Quantification of N_2O and CO_2

The impact of biochar and poultry litter alone and in combined treatments on the loss of N and C by way of greenhouse gas emission (N_2O and CO_2) were measured during 2010 and 2011 growing seasons (Table 5 and Figure 4). Poultry litter treatment produced significantly greater N_2O emission in both years; however, the N_2O emission decreased significantly when PL was combined with biochar in 2010. The decrease in N_2O emission was most likely due to the conversion of different N forms from PL by microorganism activities which occur with biochar presence. This is due to the ability for biochar to provide a habitat for microorganisms, thus increasing microbial activities. Biochar has been shown to both increase and lower nitrous oxide (N_2O) emissions [9] which seem to be related to the type of biochar used and the soil properties. In this study, the latter occurred. The N_2O emission from biochar was the same as the control. The N_2O fluxes did not increase significantly when biochar was combined with PL or fertilizer, except in 2010 (Table 5). Fertilizer treatment resulted in significantly greater N_2O emission than when it was combined with biochar. The N_2O emission factors (emission per total N applied for each treatment) and emission based on yield scale followed the same trend as N_2O emission for each treatment (Table 5). Similar to N_2O, CO_2 emission was greatest with PL treatment. In general, biochar slightly lowered the emission of N_2O and CO_2 when combined with PL or fertilizer in both years (Table 5). Cumulative N_2O and CO_2 fluxes were lower for the first 2 weeks after application (DOY 135, mid-May) and then increased drastically for the following 2 weeks until mid-June (DOY 165) (Figure 4). Poultry litter treatment produced greater (mostly significant) cumulative N_2O and CO_2 fluxes than other treatments. The cumulative fluxes were higher in 2010 than in 2011. The higher fluxes may be due to the fresh application of litter, biochar, and fertilizer in 2010 (Figure 4), along with air and soil moisture content differences in each year (Figure 5). We also speculate that the disturbance of the experimental plots from no-till grass covered plots (in order to apply biochar and planting corn) may have caused a temporary aerobic condition and increased microbial activities resulting in influxes of more N_2O and CO_2 in 2010 compared to 2011 (Figure 4, Table 5).

Table 5. Cumulative growing-season CO_2-C and N_2O-N emissions, N_2O-N emission factor, and yield-scaled N_2O emissions for each treatment in 2010 and 2011.

Treatment	CO_2-C Emissions		N_2O-N Emissions		N_2O-N Emission Factor ‡		Yield Scaled N_2O-N Emissions	
	2010	2011	2010	2011	2010	2011	2010	2011
	Mg ha^{-1}		kg ha^{-1}		%		kg N_2O-N Mg grain yield^{-1}	
Control	4.04ab †	2.80ab	4.72d	1.09c	-	-	0.73e	0.15c
Biochar (B)	3.94ab	2.61bc	5.01cd	1.36c	-	-	0.91de	0.21c
Fertilizer (F)	3.92ab	2.16cd	8.53b	3.92ab	1.70a	1.26a	1.82b	0.63b
Poultry litter (PL)	4.73a	3.19a	11.95a	4.76a	1.61a	0.82a	2.54a	0.88a
B + F	3.52b	1.89d	6.16c	3.43b	0.64b	1.05a	1.22c	0.51b
B + PL	4.04ab	2.94ab	5.83cd	4.24ab	0.25b	0.70a	1.09cd	0.71ab

† Values for each year (column) followed by the same letters are not significantly different according to Fisher's LSD (0.05) level. ‡ Calculated based on the TN application rate for each treatment.

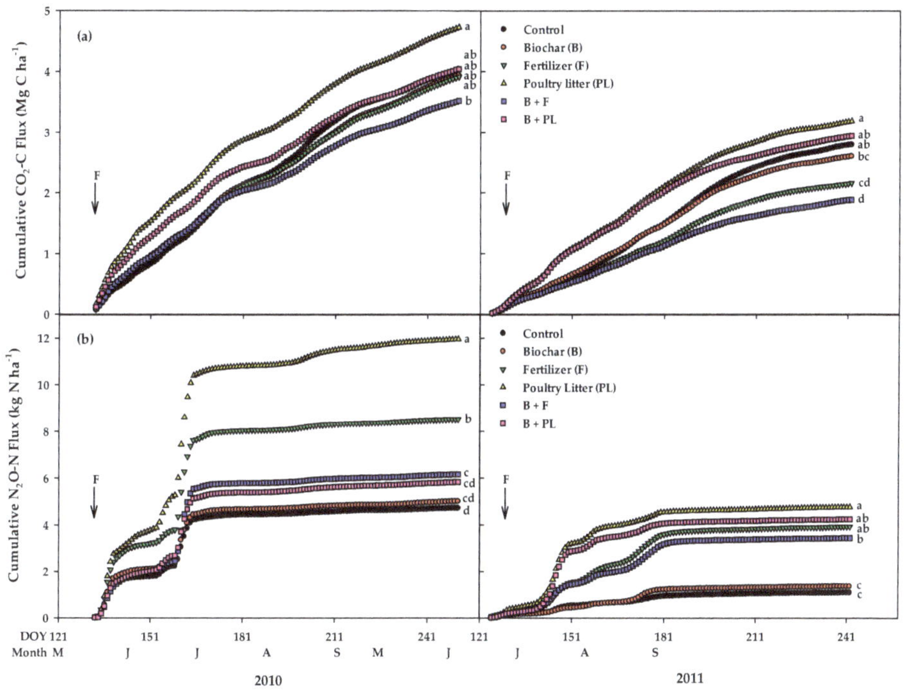

Figure 4. (**a**) Cumulative CO_2-C and (**b**) N_2O-N emissions for each treatment during the 2010 and 2011 growing seasons as a function of day of year (DOY). Cumulative values within each year for same effect followed by the same letters are not significantly different according to Fisher's LSD (0.05) level. B, biochar; F, fertilizer; PL, poultry litter.

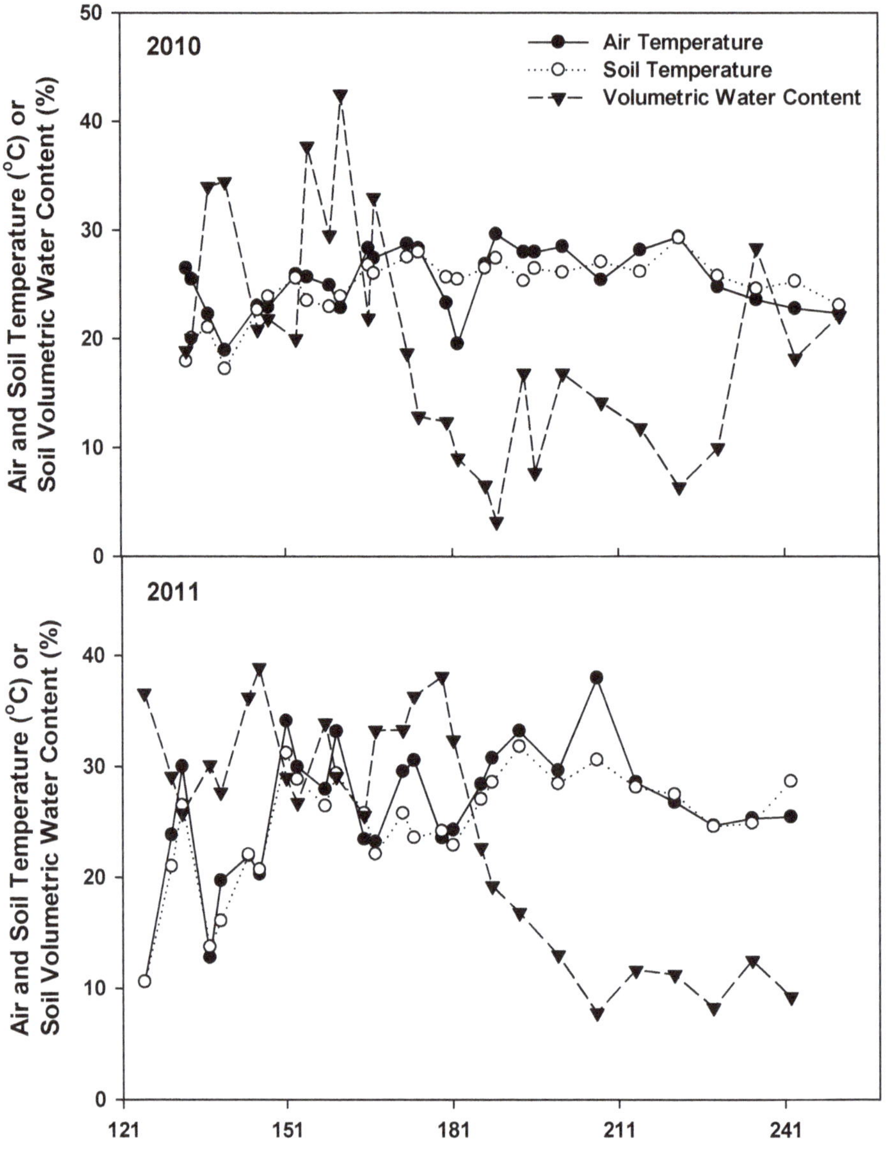

Figure 5. Air and soil temperature and soil moisture content measured during the first 2 years (2010 and 2011) of the study.

4. Conclusions

We conducted a field plot study to investigate the impact of biochar and poultry litter alone or in combination on corn biomass, grain yield, nutrient uptake, and greenhouse gas emission (N_2O and CO_2) for three growing seasons. Results indicate that biochar alone application to soil did not increase corn grain yield or N, P, and K uptake when compared to other treatments. This finding is in line with other research reports [36,37]. However, grain yield increased slightly over time from biochar application. These mixed results of delayed positive responses and even initial negative responses observed from the biochar in this study have also been reported by other research [2]. Soil pH for biochar treatment increased slightly from 4.7 (background) to 4.81 at the end of the study. Similar trends of slight increases in total C (22.4 g kg^{-1}) and total N (1.7 g kg^{-1}) observed for biochar treatments compared to background. However, we observed greater concentrations of P (7.95 mg kg^{-1}) and K (157 mg kg^{-1}) at the end of the study for biochar and biochar plus poultry litter treatments compared to soil background concentrations. Poultry litter application alone produced a significantly greater corn yield than biochar but was similar to chemical fertilizer. The N, P, and K uptake by corn grown on biochar alone applications were significantly lower than PL and fertilizer treatments. Results indicated that the addition of fertilizer or poultry litter to biochar had a positive effect on reducing N_2O and CO_2 fluxes compared to fertilizer or poultry litter application alone. In general, biochar application did not show a significant improvement in corn production parameters measured when compared to other treatments under the specific soil and environmental conditions of the study site within the three growing seasons. However, positive results were observed, such as a slight increase in grain yield in each year following biochar application and also when biochar was mixed with poultry litter or fertilizers. Therefore, more research is warranted to include different types of biochar and different rates of application under different environmental and management conditions and for longer term periods (long-term studies) to understand the impact of biochar as a soil amendment.

Author Contributions: All the authors (K.R.S., J.R.S., M.J.-B., and J.M.N.) substantially contributed to the Conceptualization, Investigation, Methodology, Data Analyses, and Preparation of this manuscript.

Funding: This work was funded by the United State, Department of Agriculture (USDA-ARS).

Conflicts of Interest: The authors declare no conflict of interest.

References

1. Lehmann, J. Bio-Energy in the black. *Front. Ecol. Environ.* **2007**, *5*, 381–387. [CrossRef]
2. Spokas, K.A.; Cantrell, K.B.; Novak, J.M.; Archer, D.A.; Ippolito, J.A.; Collins, H.P.; Boateng, A.A.; Lima, I.M.; Lamb, M.C.; McAloon, A.J.; et al. Biochar: A synthesis of its agronomic impact beyond carbon sequestration. *J. Environ. Qual.* **2012**, *41*, 973–989. [CrossRef]
3. Lehmann, J.; Abiven, S.; Kleber, M.; Pan, G.; Singh, B.P.; Sohi, S.P.; Zimmerman, A.R. Persistence of biochar in soil. In *Biochar for Environmental Management: Science, Technology and Implementation*, 2nd ed.; Lehmann, J.L., Joseph, S., Eds.; Earthscan: London, UK, 2015; pp. 235–282.
4. Fowles, M. Black carbon sequestration as an alternative to bioenergy. *Biomass Bioenergy* **2007**, *31*, 426–432. [CrossRef]
5. Glaser, B.; Lehmann, J.; Zech, W. Ameliorating physical and chemical properties of highly weathered soils in the tropics with charcoal-a review. *Biol. Fertil. Soils* **2002**, *35*, 219–230. [CrossRef]
6. Liang, B.; Lehmann, J.; Solomon, D.; Kinyangi, J.; Grossman, J.; O'Neill, B.; Skjemstad, J.O.; Theis, J.; Luizao, R.J.; Peterson, J.; et al. Black carbon increases cation exchange capacity in soils. *Soil Sci. Soc. Am. J.* **2006**, *70*, 1719–1730. [CrossRef]
7. Chan, K.Y.; Van Zwieten, L.; Meszaros, I.; Downie, A.; Joseph, S. Agronomic values of green-waste biochar as soil amendment. *Aust. J. Soil Res.* **2007**, *45*, 629–634. [CrossRef]
8. Biederman, L.A.; Harpole, W.S. Biochar and its effects on plant productivity and nutrient cycling: A meta-analysis. *GCB Bioenergy* **2013**, *5*, 202–214. [CrossRef]
9. Clough, T.J.; Condron, L.M. Biochar and the nitrogen cycle: Introduction. *J. Environ. Qual.* **2010**, *39*, 1218–1223. [CrossRef] [PubMed]
10. Atkinson, C.J.; Fitzgerald, J.D.; Hipps, N.A. Potential mechanisms for achieving agricultural benefits from biochar application to temperate soils: a review. *Plant Soil* **2010**, *337*, 1–18. [CrossRef]
11. Lehmann, J.; Rilling, M.; Thies, J.; Masiello, C.A.; Hockaday, W.C.; Crowley, D. Biochar effects on soil biota: A review. *Soil Biol. Biochem.* **2011**, *43*, 1812–1836. [CrossRef]
12. Novak, J.M.; Spokas, K.A.; Cantrell, K.B.; Ro, K.S.; Watts, D.W.; Glaz, B.; Busscher, W.J.; Hunt, P.G. Effects of biochar and hydrochars produced from lignocellulosic and animal manure on fertility of a Mollisol and Entisol. *Soil Use Manag.* **2014**, *30*, 175–181. [CrossRef]

13. Spokas, K.A.; Reicosky, D. Impacts of sixteen different biochars on soil greenhouse gas production. *Ann. Environ. Sci.* **2009**, *3*, 179–193.
14. Van Zwieten, L.; Kimber, S.; Morris, S.; Chan, K.Y.; Downie, A.; Rust, J.; Joesph, S.; Cowie, A. Effects of biochar from slow pyrolysis of papermill waste on agronomic performance and soil fertility. *Plant Soil* **2010**, *327*, 235–246. [CrossRef]
15. Novak, J.M.; Lima, I.M.; Xing, B.; Gaskin, J.W.; Steiner, C.; Das, K.C.; Ahmedna, M.; Rehrah, D.; Watts, D.W.; Busscher, W.J.; et al. Characterization of designer biochar produced at different temperatures and their effects on a loamy sand. *Soil Sci.* **2009**, *3*, 195–206.
16. Novak, J.M.; Ro, K.; Ok, Y.S.; Sigua, G.; Spokas, K.; Uchimiya, S.; Bolan, N. Biochars multifunctional role as a novel technology in the agricultural, environmental, and industrial sectors. *Chemosphere* **2016**, *142*, 1–3. [CrossRef] [PubMed]
17. Jeffery, S.; Verheijen, F.G.; Van der Velde, M.; Bastos, A.C. A quantitative review of the effects of biochar application to soils on crop productivity using meta-analysis. *Agric. Ecosyst. Environ.* **2011**, *144*, 175–187. [CrossRef]
18. Crane-Droesch, A.; Abiven, S.; Jeffery, S.; Torn, M.S. Heterogeneous global crop yield response to biochar: A meta-regression analysis. *Environ. Res. Lett.* **2013**, *8*, 1–8. [CrossRef]
19. Ronsee, F.; Van Hecke, S.; Dickinson, D.; Prins, W. Production and characterization of slow pyrolysis biochar: Influence of feedstock type and pyrolysis conditions. *GCB Bioenergy* **2013**, *5*, 104–115. [CrossRef]
20. Hass, A.; Gonzalez, J.M.; Lima, I.M.; Godwin, H.W.; Halvorson, J.J.; Boyer, D.G. Chicken manure biochar as liming and nutrient source for acid Appalachian soil. *J. Environ. Qual.* **2012**, *41*, 1096–1106. [CrossRef]
21. Wang, L.; Butterly, C.R.; Wang, Y.; Herath, H.M.; Xi, Y.G.; Xiao, X.J. Effect of crop residue biochar on soil acidity amelioration in strongly acidic tea garden soils. *Soil Use Manag.* **2014**, *40*, 119–128. [CrossRef]
22. Keeney, D.R.; Nelson, D.W. Nitrogen-inorganic forms. In *Methods of Soil Analysis, Part 2*, 2nd ed.; Page, A.L., Miller, R.H., Keeney, D.R., Eds.; ASA and SSSA: Madison, WI, USA, 1982; pp. 643–698.
23. Mehlich, A. Mehlich 3 soil extractant: A modification of Mehlich-2 extractant. *Commun. Soil Sci. Plant Anal.* **1984**, *15*, 1409–1416. [CrossRef]
24. Novak, J.M.; Busscher, W.J.; Laird, D.L.; Ahmedna, M.; Watts, D.W.; Niandou, M.A. Impact of biochar on fertility of a southeastern Coastal Plain soil. *Soil Sci.* **2009**, *74*, 105–112. [CrossRef]
25. Miller, R.O. High-temperature oxidation: Dry ashing. In *Referenced Methods for Plant Analysis*; Kalra, V.P., Ed.; Handbook of. CRC Press: Boca Raton, FL, USA, 1998; pp. 53–56.
26. Livingston, G.P.; Hutchinson, G.L. Enclosure-based measurement of trace gas exchange: Applications and sources of error. In *Biogenic Trace Gases: Measuring Emissions from Soil and Water*; Matson, P.A., Harriss, R.C., Eds.; Blackwell Sci. Ltd.: London, UK, 1995.
27. Mosier, A.R.; Halvorson, A.D.; Reule, C.A.; Liu, X.J. Net global warming potential and greenhouse gas intensity in irrigated cropping systems in northeastern Colorado. *J. Environ. Qual.* **2006**, *35*, 1584–1598. [CrossRef] [PubMed]
28. Hutchinson, G.L.; Mosier, A.R. Improved soil cover method for field measurements of nitrous oxide fluxes. *Soil Sci. Soc. Am. J.* **1981**, *45*, 311–316. [CrossRef]
29. Yamato, M.; Okimori, Y.; Wibowo, I.F.; Anshori, S.; Ogawa, M. Effects of the application of charred bark of Acaciamangium on the yield of maize, cowpea and peanut and soil chemical properties in south Sumatra, Indonesia. *Soil Sci. Plant Nut.* **2006**, *52*, 489–495. [CrossRef]
30. Steiner, C.; Teixeira, W.G.; Lehmann, J.; Nehls, T.; Macedo, J.; Blum, W.H.; Zech, W. Long term effect of manure, charcoal and mineral fertilization on crop production and fertility on a highly weathered Central Amazonian upland soil. *Plant Soil* **2007**, *291*, 275–290. [CrossRef]
31. Mau, A.E.; Utami, S.R. Effects of biochar amendment and arbuscular mycorrhizal fungi inoculation on availability of soil phosphorus and growth of maize. *J. Degr. Min. Lands Manag.* **2014**, *1*, 69–74.
32. Ahmad, M.A.; Asghar, H.N.; Saleem, M.; Khan, M.Y.; Zahir, Z.A. Synergistic effect of rhizobia and biochar on growth and physiology of maize. *Agron. J.* **2015**, *107*, 2327–2334. [CrossRef]
33. Adejumo, S.A.; Owolabi, M.O.; Odesola, I.F. Agro-physiologic effects of compost and biochar produced at different temperatures on growth, photosynthetic pigment and micronutrients uptake of maize crop. *Afr. J. Agric. Res.* **2015**, *11*, 661–673. [CrossRef]
34. Eghball, B.; Wienhold, B.J.; Gilley, J.E.; Eigenberg, R.A. Mineralization of manure nutrients. *J. Soil Water Conserv.* **2002**, *57*, 470–478.

35. Motavalli, P.P.; Kelling, K.A.; Converse, J.C. First-year nutrient availability from injected dairy manure. *J. Environ. Qual.* **1998**, *18*, 180–185. [CrossRef]
36. Lentz, R.D.; Ippolito, J.A.; Spokas, K.A. Biochar and Manure Effects on Net Nitrogen Mineralization and Greenhouse Gas Emissions from Calcareous Soil under Corn. *Soil Sci. Soc. Am. J.* **2014**, *78*, 1641–1655. [CrossRef]
37. Laird, D.A.; Novak, J.M.; Collins, H.P.; Ippolito, J.A.; Karlen, D.L.; Lentz, R.D.; Sistani, K.R.; Spokas, K.; van Pelt, R.S. Multi-year and multi-location soil quality and crop biomass yield responses to hardwood fast pyrolysis biochar. *Geoderma* **2017**, *289*, 46–53. [CrossRef]

environments

MDPI

Article

Fertilizer Efficacy of Poultry Litter Ash Blended with Lime or Gypsum as Fillers

Philip J. Bauer, Ariel A. Szogi * and Paul D. Shumaker

United States Department of Agriculture-Agricultural Research Service, Coastal Plains, Soil, Water and Plant Research Center, 2611 W. Lucas St., Florence, SC 29501, USA; lbauer@sc.rr.com (P.J.B.); paul.shumaker@ars.usda.gov (P.D.S.)

* Correspondence: ariel.szogi@ars.usda.gov; Tel.: +1-843-669-5203

Received: 2 April 2019; Accepted: 29 April 2019; Published: 1 May 2019

Abstract: Ash from power plants that incinerate poultry litter has fertilizer value, but research is lacking on optimal land application methodologies. Experiments were conducted to evaluate calcitic lime and flue gas desulfurization gypsum (FGDG) as potential fillers for poultry litter ash land applications. The ash had phosphorus (P) and potassium (K) contents of 68 and 59 g kg^{-1}, respectively. Soil extractable P and K were measured in an incubation pot study, comparing calcitic lime to FGDG at filler/ash ratios of 1:3, 1:2, 1:1, 2:1, and 3:1. After one month, soils were sampled and annual ryegrass (*Lolium multiflorum* Lam.) seeds were planted to investigate how plant growth and uptake of P and K were influenced by the fillers. Application of ash alone or with fillers increased soil extractable P and K levels above unamended controls by 100% and 70%, respectively. Filler materials did not affect biomass or P and K concentration of the ryegrass. A field study with a commercial spinner disc fertilizer applicator was conducted to compare application uniformity of ash alone and filler/ash blends. Overall, test data suggested that uniform distribution of ash alone or with fillers is feasible in field applications using a commercial fertilizer spreader.

Keywords: poultry litter ash; fertilizer filler; phosphorus; potassium; flue gas desulfurization gypsum; calcitic lime

1. Introduction

The intensive production of poultry generates large amounts of spent litter, most of which is applied to agricultural land as a nutrient source in forage and crop production [1]. Repeated land applications of spent poultry litter have resulted in many fields containing nutrient levels above the assimilative capacity of soils [2]. In turn, these high soil nutrient levels cause great concern because of the environmental consequences associated with air and water quality [3,4]. As an alternative to land application, poultry litter incineration is being adopted in Europe and the United States to produce energy and reduce the volume of disposed poultry litter [1,2,5]. The ash from incinerated poultry litter has fertilizer value with high concentrations of plant nutrients, such as phosphorus (P) and potassium (K). Recycling these nutrients is essential to close the nutrient loop in food systems, given that both P and K are limited mined resources being depleted by global demand in production agriculture [6,7]. By concentrating nutrients such as P and K in the ash, it is more economically feasible to transport them to distant croplands with nutrient deficient soils. However, research is needed for environmentally safe and uniform field applications of poultry litter ash products.

Nutrient content of ash generated from poultry litter can be widely variable, due to the type of poultry production (i.e., broilers, layers, bird species), nutrient content of the rations fed to the birds, type of bedding material, number of birds in a flock, number of flocks, and incineration process conditions [2,5,8]. Nevertheless, poultry litter ash usually contains substantial amounts of P and K. For example, poultry litter ash from a commercial farm in South Carolina contained minimal contents of nitrogen (N) but had 9.1% P_2O_5 and 7.9% K_2O [9]. Thus, poultry litter ash has generally been reported as a satisfactory source of nutrients for crops [10–12]. While as much as 20% of the P in raw poultry litter is readily water-soluble [13], P in poultry litter ash is relatively insoluble in water [8,14]. In a P fractionation study of poultry litter ash, Codling [15] found water-extractable P to be about 1.5% of the total P. Potassium, on the other hand, is highly water-soluble in raw poultry litter and it remains water-soluble even after poultry litter is converted into ash [14,16].

Fertilizers usually contain two types of ingredients: Active and inactive. The active ingredients are the plant macro- and micro-nutrients. The inactive ingredients, also called fillers, may include sand, granular limestone, or sawdust. Filler materials are commonly added to commercial fertilizer blends, so nutrients are evenly applied in amounts based on soil test results. Two readily available potential fillers for the Southeast USA are calcitic limestone and flue gas desulfurization gypsum (FGDG). Calcitic limestone is used to neutralize acid soils, especially in fields not requiring magnesium [17]. In 2016, about 670 thousand tons of FGDG were used in agriculture as a source of calcium and sulfur and to improve the soil's physical properties [18,19]. In addition, fillers may improve the distribution uniformity of the spreading of ash on agricultural fields, as concerns have been raised about the need to uniformly apply poultry litter ash [9].

Codling et al. [20] evaluated how broiler litter ash with and without FGDG affected peanut yield and nutrient uptake, but in their experiment, the ash was mixed in the soil and incubated for three weeks before planting peanuts and surface applying FGDG. Apart from the Codling et al. [20] study, we are unaware of other published research evaluating the availability of P and K in poultry litter ash when applied with either calcitic lime or FGDG. Use of poultry litter ash directly from the incinerator with minimal processing can make it cost effective and attractive as a substitute for commercial fertilizers. Except for blending with lime or FGDG, our study did not include additional processing, common in commercial fertilizer production (crushing, extrusion, granulation, etc.). Meanwhile, farmers use both conventional and conservation tillage. Placement of the ash and fillers (mixed with soil or left on the surface) needs to be evaluated for both management systems for the efficacy of ash–filler blends to supply nutrients. The goal of this research was to evaluate fertilizer effectiveness of poultry litter ash when applied in blends with calcitic lime and FGDG. We conducted a controlled environment study with the objective of determining whether placement, filler material, and filler/ash ratios influence soil-extractable P and K, plant growth, and plant uptake of those nutrients. In addition, we conducted a field study to determine if blends of ash with calcitic lime or FGDG improved uniformity of application over ash by itself with commercial fertilizer application equipment.

2. Materials and Methods

2.1. Characterization of Ash, Calcitic Lime, and FGDG

The poultry litter ash used in the soil and plant response study was turkey litter ash from a power plant. It was provided by Carolina Eastern, Inc. (Charleston, SC, USA). The FGDG was also provided by Carolina Eastern, Inc. and was from a coal-fueled power plant. Calcitic lime was from a mine near Loris, SC, USA and was provided by Wake Stone Corporation, Knightdale, NC, USA. Elemental analysis of the ash, lime, and FGDG was determined by digesting the materials in nitric acid with peroxide (EPA 3050B), using a block digester [21], followed by quantifying elements in the digest using inductively coupled plasma optical emission spectroscopy (ICP-OES).

2.2. Plant-Available P of Ash and Ash–Filler Blends

Duplicate samples of ash and ash blended with fillers (calcitic lime and FGDG) at different ratios (3:1, 2:1, 1:1, 1:2, and 1:3) on a mass basis were analyzed for water-soluble P, citrate-insoluble P, and citrate-soluble P as $\%P_2O_5$, according to AOAC Official Methods 958.01, 977.01, 963.03 B(a) [22]. The "plant-available P" was determined using the AOAC "available P" test for fertilizing materials and it is the citrate-soluble P (which includes the water-soluble P fraction) [22]. Total N in the ash was determined by combustion with a Leco TruSpec CN analyzer (Leco Corp., St. Joseph, MI, USA).

2.3. Soil and Plant Resposnse

Soil was collected from the surface, 15 cm of a Norfolk loamy sand (fine-loamy, siliceous, thermic Typic Kandiudults) at Clemson University's Pee Dee Research and Education Center (Florence, SC, USA). The collected soil was spread out on greenhouse benches to dry, prior to conducting the experiment. After drying, the soil was passed through a 6.35-mm screen to remove roots and large soil particles.

A controlled environment study was conducted with the following treatments: Unamended control, 100% ash, and ash supplied as 75, 66, 50, 33, and 25% of mixtures with either calcitic lime or FGDG. The selected ash levels matched the filler/ash ratios of 3:1, 2:1, 1:1, 1:2, and 1:3. In all treatment combinations (except the unfertilized control), the rates of P and K applied to the soil were the same. The 100% ash treatment was added to the soil at a ratio of 0.9 g ash to 1.0 kg soil. Prior to soil application, calcitic lime and FGDG were first blended with ash for each filler/ash treatment that resulted in application rates of 0.3, 0.6, 0.9, 1.8, and 2.7 g filler per kg of soil. The study was conducted using 20-cm diameter pots in which the soil depth was 12 cm. The ash, FGDG, and calcitic lime were all applied on an air-dry mass basis, either incorporated or left on the soil surface. There were three replicates in the experiment and the experiment was conducted twice. Pots were watered to 100 g kg^{-1} soil water content, covered with newsprint, and stored in a room with no environmental control for 30 days during the summer. During storage, pots were monitored and a small amount of water was added if the soil surface appeared dry.

After 30 days, a 1-cm diameter cork borer was used to sample the soil in the pots to a depth of 10 cm. Six cores were collected from each pot, homogonized, and dried at 60 °C for three days. Following drying, soil pH (1:1, soil to water) was measured and plant-available P and K in soil samples were quantified by ICP-OES in Mehlich-1 extracts [23]. The plant-available P and K in soil extracts hereafter are called "available soil P and K", respectively.

After soil sampling, the pots were moved to a greenhouse and the holes made by sampling were filled with unamended soil. Ryegrass seed (1.11 g per pot) was placed on the soil surface of each pot and covered with dry unamended soil. The pots were watered as needed with tap water for the duration of the experiment. Three harvests of plant shoot tissues were made by cutting plants 2.5 cm above the soil surface three, five, and seven weeks after planting. A small amount of N (1.0 g of NH_4NO_3) was added to each pot following the first two harvests. Plant tissue samples were dried at 60 °C for three days and then weighed and ground. Total P and K concentrations in digested plant tissues were quantified using the same method used for ash characterization [21].

2.4. Spreading Uniformity Test

The spreading uniformity of ash, ash with calcitic lime, and ash with FGDG was evaluated using a commercial spinner disc fertilizer/lime spreading truck. The source of the ash supplied by Carolina Eastern Inc. was from a power plant in the region that co-incinerates wood and poultry litter. This test consisted of three treatments: Ash alone, ash mixed with calcitic lime, and ash mixed with FGDG. It was conducted under low wind speeds (<1.8 m s^{-1}) on a crop field under conservation tillage with less than 2% slope.

A front-end loader and a belt elevator conveyor were used to load all materials into the spreading truck. Lime and FGDG were mixed with ash to make 1:4 (filler/ash) blends. This blend ratio was requested by the supplier to further reduce the filler/ash ratio, according to our laboratory tests. To make the blends, four front-end loads of ash followed by a load of filler material were loaded into the truck with a belt elevator conveyor. This was repeated until the truck contained 16 loads of ash and four loads of lime or FGDG. The filler/ash mix in the truck was then emptied onto the ground, into a pile, repeatedly scooped and dumped with the front-end loader, and then reloaded into the truck using the belt elevator conveyor. Bulk densities of the ash and the mixtures were estimated using a hand-held Fertilizer Density Scale (Berckes Mfg., Canby, MN, USA). The bulk density of the ash was approximately 481 kg m^{-3} while the bulk density of both the lime/ash and FGDG/ash mixtures were approximately 641 kg m^{-3}. These density values were used to adjust settings in the truck so all application rates were about 2242 kg ha^{-1}. Duplicate samples of the three treatments were collected for gravimetric water content determinations. Water content of the ash without filler was 112 g kg^{-1}, the FGDG/ash blend was 111 g kg^{-1}, and the lime/ash blend was 88 g kg^{-1}. Particle densities were determined by pouring a known mass of product into a known volume of distilled water and immediately recording the new total volume. The particle density was calculated as the dry mass of the product divided by the displaced volume. Particle densities for ash alone and mixtures of lime/ash and FGDG/ash were 1.93, 2.02, and 1.96 g mL^{-1}, respectively. Particle size distribution of ash, calcitic lime, and FGDG alone and mixtures of lime/ash and FGDG/ash were determined using a sieve shaker to pass a known mass through American Society for Testing and Materials (ASTM) sieve Nos. 1/2, 5/16, 5, 10, 18, and 35. (Table 1).

Table 1. Particle size distribution of ash alone, lime, and flue gas desulfurization gypsum (FGDG), and mixtures of 1:4 lime/ash and 1:4 FGDG/ash used in the field spreading uniformity experiment. Data are the mean of four replicate samples.

Particle Size	Ash	Lime	FGDG	Lime/Ash	FGDG/Ash
mm	Percent finer by weight				
>12.5	3.1	0.0	0.0	4.2	2.2
8.0–12.7	3.9	0.0	1.0	3.5	2.9
4.0–8.0	12.9	0.6	4.7	12.3	12.8
2.0–4.0	17.3	0.7	3.7	13.9	14.7
1.0–2.0	18.3	1.7	1.6	12.8	13.1
0.5–1.0	16.6	14.3	0.5	15.5	11.9
<0.5	27.9	82.7	88.5	37.7	42.4

The spreading uniformity of the three treatments was evaluated using a catch-pan method, typically used to calibrate spreader applicators [24]. Catch pans were placed 1.5 m apart along a line perpendicular to the direction of travel of the spreader truck. The spreader application test method was done such that the spreader truck was driven next to the pan at the end of the line when performing the test. The distribution of the spread of the ash and filler/ash blends was evaluated on each side (right on right and left on left swaths) of the spreader application [24]. Evaluation of material spread on both sides of the spreader was considered one replication. The total mass of the materials caught in each individual evaluation (right and left) was determined and the mass of material in each catch pan was converted to a percentage of the total. Treatments were replicated twice.

2.5. Data Analysis

Data were analyzed using SAS version 9.4. All data from the controlled environment study were analyzed in two ways. First, to determine if interactions occurred among placement (placement was not a part of the analysis for plant available P), filler material, and filler ratio, an analysis of variance (ANOVA) was conducted, excluding the data for the unamended control and ash alone. Means of significant interactions were separated using pairwise comparisons. Second, main effect means were compared using all the data by conducting an ANOVA and computing single degree of freedom contrasts. The contrasts compared means of ash alone to the control, calcitic lime to the control, FGDG to the control, calcitic lime to ash alone, FGDG to ash alone, and calcitic lime to FGDG. Sources of variation and contrasts were considered significant when probability of >F values were ≤0.05. An ANOVA was conducted to determine the effect of ash alone and the different filler/ash ratio combinations on ryegrass biomass and P and K content in plant tissue. For the spreader uniformity study, standard deviation and percent coefficient of variation were calculated for each treatment at each swath spacing from the spreader truck [24].

3. Results and Discussion

3.1. Poultry Litter Ash, Calcitic Lime, and FGDG Characterization

The turkey litter ash used to evaluate soil and plant response had high amounts of Ca, P, and K (Table 2), which is typical of poultry litter ashes [16]. It had P and K contents equivalent to 15.5% P_2O_5 and 7.0% K_2O, but almost no N (0.6%) because incineration causes almost all the N in feedstocks to be converted to N_2 and nitric oxide gases [9,25]. Both calcitic lime and FGDG had, as expected, high amounts of Ca and FGDG had a high amount of S, while the P and K content of the filler materials was very low (Table 2). The concentration of Cu and Zn in poultry litter ash is of agronomic concern because of the risk of accumulation in soils at toxic levels for plants. In our study, a soil application rate of 0.9 g ash kg^{-1} was equivalent to applying 1.0 mg Cu kg^{-1} and 0.9 mg Zn kg^{-1}. These Cu and Zn rates are well below the total Cu and Zn concentrations of 8.5 mg Cu kg^{-1} and 20.1 mg Zn kg^{-1} found in sandy soils of the U.S. Coastal Plain region, impacted by long-term application of swine manure [26]. Concentrations of other plant nutrients in the ash and the two filler materials are shown in Table 2.

Table 2. Plant nutrient composition of the poultry litter ash, calcitic lime, flue gas desulfurization gypsum (FGDG) used in the greenhouse experiment.

Plant Nutrient	Ash	Lime	FGDG
P (g kg^{-1})	68	0.1	0.04
K (g kg^{-1})	59	0.3	0.4
Ca (g kg^{-1})	134	396	250
Mg (g kg^{-1})	13	3	0.6
S (g kg^{-1})	8	7	192
Cu (mg kg^{-1})	1151	BD [1]	BD
Fe (mg kg^{-1})	4827	2847	541
Mn (mg kg^{-1})	1084	40	BD
Mo (mg kg^{-1})	12	BD	BD
Zn (mg kg^{-1})	797	BD	BD

[1] BD indicates that nutrient was below detection level.

3.2. Plant-Available P in Ash Material

Averaged over all treatment combinations, 44% of the total P in the ash was plant-available, according to the AOAC citrate-soluble test [22]. Previously, Clarholm [27] found that only 20% of the total P in granulated wood ash was extractable with ammonium acetate, while Codling [15] found that hydrochloric acid extractable P in poultry litter ash was 82% of total P. Table 3 shows that, averaged over all filler/ash ratios, the two filler materials did not significantly change plant-available P from ash alone (100% ash). For ash blended with lime, none of the treatment combinations differed from the lime 1:2 filler/ash ratio, which was numerically similar to the plant-available P of ash alone. The lime 1:3 filler/ash ratio had the lowest percentage of plant-available P (34.8%), while the FGDG 1:3 filler/ash ratio had the highest (48.5%). Averaged across ratios, the ash blended with lime had approximately 8% more plant-available P than ash blended with FGDG. A significant filler × ratio interaction supported the observations that increasing amounts of lime enhanced percentage of P availability, while increasing amounts of FGDG diminished percentage of P availability (Table 3).

Table 3. Percentage of the P in ash that was plant-available as affected by calcitic lime and flue gas desulfurization gypsum (FGDG).

Filler	Ratio	Ash	Plant Available P
	Filler/Ash	Percentage	Percentage
Lime	1:3	75	34.8de [1]
	1:2	66	46.9abcde
	1:1	50	55.7a
	2:1	33	52.0ab
	3:1	25	51.3abc
FGDG	1:3	75	48.5abcd
	1:2	66	37.0cde
	1:1	50	33.7e
	2:1	33	41.3bcde
	3:1	25	39.3bcde
$p > F$ [2]			0.03
Means Over Ratio			
Lime			48.2
FGDG			40.0
Ash			46.5
Contrast Comparisons of Means			$p > F$
Lime vs. Ash			0.74
FGDG vs. Ash			0.20
Lime vs. FGDG			0.01

[1] Means followed by the same letter are not significantly different according to least square difference (LSD 0.05).
[2] Probability of a greater F value of the filler × ratio interaction.

In previous work, Codling et al. [15] reported that poultry litter ash contains very little to no water-soluble P. Similarly, the ash materials in our study had extremely low water-soluble P (0.37% P_2O_5). Gypsum mixed with animal manure has been found to reduce water-soluble P [28–30], which may result from Ca in the gypsum binding with the water-soluble P to form water-insoluble calcium phosphate [31]. Watts and Torbert [32] found that gypsum applied to grass buffer strips downslope from a poultry litter application reduced soluble P in runoff. Furthermore, Endale et al. [33] found that FGDG applied with poultry litter reduced soluble P in runoff in one of two years. However, the low amount of water-soluble P in poultry litter ash suggests that such binding effects of gypsum with soluble water P with this P fertilizer source would be negligible.

3.3. Plant Available P and K in Soil

Lime and FGDG as fillers did not affect how available soil P and K levels responded to ash placement. No interactions occurred involving placement and filler material or placement and filler ratio. Similarly, the filler material × filler/ash ratio interaction was not significant for either available soil P or K ($p > F = 0.52$ for P and 0.09 for K; Table 4). As expected, application of ash alone increased available soil P and K in soil above levels in the controls (Table 4). Ash left on the surface resulted in an available soil P level of 112 mg kg^{-1}, while ash mixed into the soil had 99 mg kg^{-1} available soil P ($p = 0.036$). Similarly, available soil K was 150 mg kg^{-1} when ash was surface-applied and 129 mg kg^{-1} when the ash was incorporated into the soil ($p < 0.001$). Higher available soil P and K levels for soil where ash was left on the surface is not surprising. Mixing the ash with soil would have distributed the nutrients throughout the pots. Since only the top 10 cm of the pots were sampled, more of the nutrients were in the sampling area for the pots with ash spread on the surface.

For available soil P, the ash alone was similar to both filler/ash blends. For available soil K, ash alone was similar to the FGDG/ash blends, but the lime/ash blends were somewhat higher than both ash alone and the FGDG/ash blends (Table 4). It is not clear why using calcitic lime as filler increased available soil K above levels in the soil amended with ash alone or the FGDG/ash blends, since both filler materials had low K concentrations (Table 2).

Table 4. Soil Available P and K (Mehlich-1) as affected by poultry litter ash amendment and filler/ash ratio. Soil was collected from pots after a 30-day incubation period.

Filler	Ratio	P	K
	Filler/Ash	mg kg^{-1}	mg kg^{-1}
Lime	1:3	97	131
	1:2	102	143
	1:1	99	139
	2:1	103	142
	3:1	136	171
FGDG	1:3	115	138
	1:2	103	141
	1:1	109	131
	2:1	90	131
	3:1	123	140
$p > F$ [1]		0.52 ns	0.09 ns
Means Over Ratios			
Lime		108	145
FGDG		108	136
Ash		86	128
Control		48	80
Contrast Comparisons of Means		$p > F$	
Ash vs. Control		0.036	<0.001
Lime vs. Control		<0.001	<0.001
FGDG vs. Control		<0.001	<0.001
Lime vs. Ash		0.06	0.02
FGDG vs. Ash		0.06	0.26
Lime vs. FGDG		0.95	0.03

[1] Probability of a greater F value of the filler × ratio interaction; $p > 0.05$; ns = non-significant difference.

3.4. Soil pH

Wood ash can be used as a liming material to neutralize acid soils. Adotey et al. [34] recently compared wood ash to two commercial liming products and found similar soil pH changes when wood ash rates were normalized for $CaCO_3$. The application rate of ash in our study corresponded to 168 kg ha^{-1} of total P_2O_5 (based on area of the top of the pots). Since plant-available P was 46.5% of total P, the application rate corresponded to 78 kg ha^{-1} of plant-available P_2O_5, which is a typical P application rate. This rate slightly raised the soil pH from 5.1 for the unamended control to 5.3 for the soil amended with ash alone and the FGDG/ash blends. As expected, FGDG did not impact soil pH, whereas ash blended with calcitic lime significantly increased soil pH. For lime/ash blend treatments, soil pH ranged from 5.8 for the 1:3 filler/ash ratio to 6.3 for the 3:1 filler/ash ratio. Applying poultry litter ash at high rates for liming purposes would result in excess P application, as discussed by Chastain et al [9]. However, blending poultry litter ash with calcitic limestone could simultaneously add P to the soil system and adjust soil pH to between 6.0 and 6.5 to favor P dissolution and availability to plants [35], thus providing a fertilizer application with recommended amounts of nutrients.

3.5. Ryegrass Biomass and P and K Uptake

Filler material and filler/ash ratio did not affect biomass or P and K concentration of the ryegrass (Figure 1). Analysis of variance results indicate that none of the filler/ratio treatment combinations differ from ash alone. These results suggest that the addition of either calcitic lime or FGDG as filler materials will not adversely affect plant uptake of P and K from soil. Others have found poultry litter ash to be effective in providing P to plants [10,36–38], while the high level of water-soluble K in ash [16] suggests it is readily plant available. Lack of response to ash application for P and K in the biomass of the ryegrass in this study was likely due to the adequate amounts of soil P and K concentrations in the soil used (48 P mg kg^{-1} and 80 mg K kg^{-1} in the unamended soil control; Table 4) [23], and the relatively short duration (seven weeks) of the plant biomass experiments.

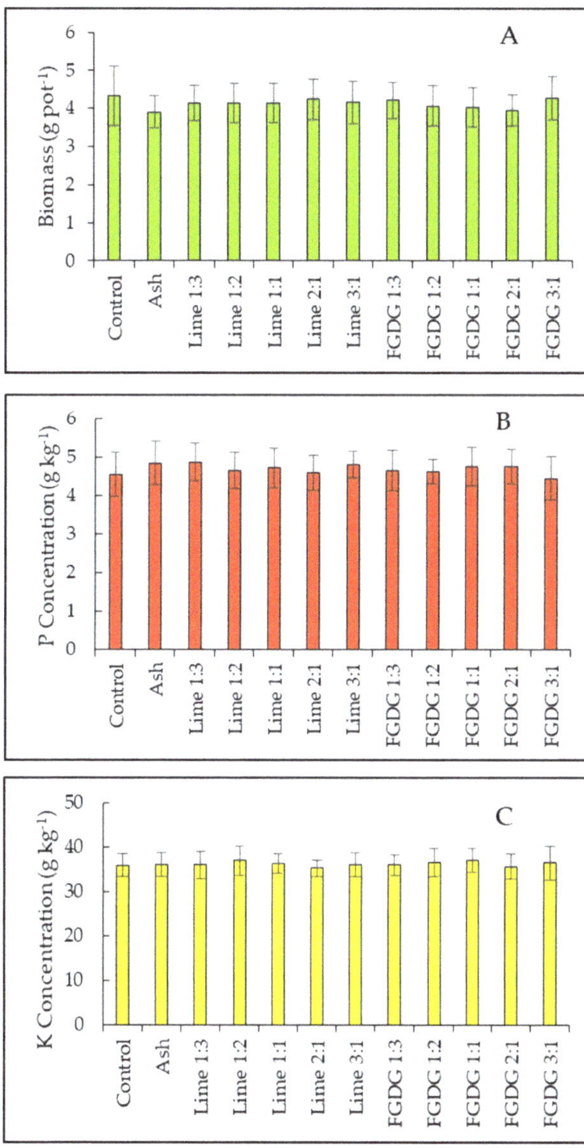

Figure 1. Effect of calcitic lime and flue gas desulfurization gypsum (FGDG) as fertilizer fillers for poultry litter ash on ryegrass biomass (**A**), P concentration in ryegrass (**B**), and K concentration in ryegrass (**C**). Error bars indicate ± one standard error of means.

3.6. Spreading Uniformity

The distribution pattern of poultry litter ash spread with and without the two filler materials is shown in Figure 2. Because of its low density, we expected application of ash alone to be irregular [39]. Surprisingly, spread of ash alone was quite uniform in the percentage of total weight within a 6.1-m swath distance from the truck (Figure 2A). Only a small amount of ash was caught in the catch pans beyond this swath width. Consistent with this distribution pattern, coefficients of vatiation (CV) values of spread uniformity for ash alone were in the range of 10 to 38% within the 6.1-m swath (Figure 2B). Even though ash alone particles have irregular shapes and a wide particle size distribution (Table 1), the CV values in our study were somewhat similar to average CV uniformity values of granular fertilizer applications (12 to 31%) [40]. Adding either filler did not substantially improve the distribution pattern of ash application but increased the swath distance by 1.5 m. At 7.6 m distance, the catch pans recovered a substantial portion of the ash material with fillers (8–10% of total weight captured in the pans). Lime/ash and FGDG/ash blends had coefficients of variation for the 6.1-m swath distance in the range of 28 to 44% and 20 to 61%, respectively (Figure 2B).

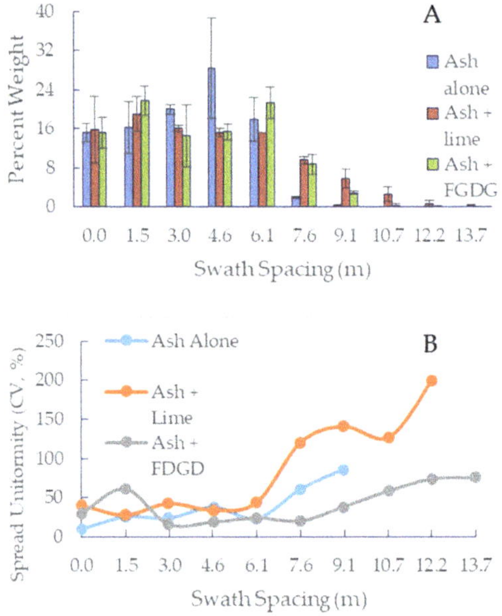

Figure 2. Spread pattern of the fertilizer applicator truck for poultry litter ash and ash mixed with calcitic lime or flue gas desulfurization gypsum (FGDG): (**A**) Average weight percent of the products at each swath spacing with respect to total weight caught in the pans; (**B**) Coefficient of variation (CV%) of spread uniformity. Swath spacing in the x-axis represents the catch pans placed 1.5 m apart along a line perpendicular to direction of travel. Error bars indicate ± one standard deviation.

In our study, the ash, calcitic lime, and FGDG were transported to the field and stored in open piles before the tests were conducted. Wetting the ash before being transported is a common practice necessary for dust control [41]. In addition, it rained between the times when the materials were delivered and when the spreading uniformity field test was performed. The wetted ash (which had a water content of about 100 g kg^{-1}) resulted in a better distribution when it was applied alone than could be expected from the spread of dry ash alone, directly from an incinerator. The addition of both fillers to wetted ash lumped some of the ash and filler blended materials into larger particle sizes than those reported in Table 1. Given that larger particles are thrown further by spinner spreaders than smaller particles [40], it explains the spreading of some particles of the lime/ash and FGDG/ash blends beyond the 7.6-swath distance. The moisture content of the products was a variable not controlled in our spreading test. Therefore, research on determining acceptable moisture contents that ensure economic transportation and optimal spreading is needed. Overall, our data suggest that spreading of the ash with commercial spinner disc applicators alone or blended with calcitic lime or FGDG as fillers is feasible using commercial fertilizer spreaders. Further evaluations of the spreading distribution of poultry litter ash both with and without fillers should be conducted at various moisture contents with varying wind speed and direction, along with fugitive dust collection studies.

4. Conclusions

Incineration of poultry litter is being used both to produce energy in power plants and as a method of waste handling and treatment in areas with high concentrations of poultry production. With its relatively high concentration of plant nutrients, poultry litter ash is a power plant byproduct with potential use as fertilizer. However, environmental concerns exist about the need to uniformly land-apply poultry litter ash. In this research, we evaluated if blending ash with two potential filler materials affected soil and plant parameters and their possible impact on field application patterns with a spinner disc applicator. The two potential fillers evaluated, calcitic lime and FGDG, appear to be appropriate filler materials for land-applying the ash. Neither of these materials negatively affected plant available concentrations of P and K in the soil, nor ryegrass biomass and plant P and K concentrations. The uniform distribution of poultry litter ash in field application using a commercial fertilizer spreader is feasible for ash alone, ash blended with calcitic lime, and ash blended with FGDG. More field testing on the potential of these fillers to enhance poultry litter ash application appears warranted, especially field studies with varying moisture contents of the products, wind speed, and wind direction that evaluate how these fillers may affect fugitive dust during application.

Author Contributions: Conceptualization, P.J.B.; Data curation, P.D.S.; Formal analysis, P.J.B. and A.A.S.; Investigation, P.J.B.; Methodology, P.J.B., A.A.S., and P.D.S.; Writing—original draft, P.J.B. and A.A.S.; Writing—review & editing, A.A.S. and P.D.S.

Funding: This research received no external funding.

Acknowledgments: This research was part of USDA-ARS National Programs 211 Water Availability and Watershed Management, Project 6082-13000-010-00D; and 212 Soil and Air ARS Project 6082-12630-001-00D. The authors thank Mr. Thomas Kemp and Carolina Eastern, Inc. for the ash, lime, and FGDG used in this research, the field equipment, and labor provided to conduct the spreader study. Mention of trade names or commercial products in this article is solely for the purpose of providing specific information and does not imply recommendation or endorsement by the U.S. Department of Agriculture.

Conflicts of Interest: The authors declare no conflict of interest.

References

1. Bolan, N.S.; Szogi, A.A.; Chuasavathi, T.; Seshadri, B.; Rothrock, M.J.; Panneerselvam, P. Uses and management of poultry litter. *World Poultry Sci. J.* **2010**, *66*, 673–698. [CrossRef]
2. Szogi, A.A.; Vanotti, M.B.; Ro, K.S. Methods for treatment of animal manures to reduce nutrient pollution prior to soil application. *Curr. Pollut. Rep.* **2015**, *1*, 47–56. [CrossRef]
3. Kaise, D.E.; Mallarino, A.P.; Haq, M.U. Runoff phosphorus loss immediately after poultry manure application as influenced by application rate and tillage. *J. Environ. Qual.* **2009**, *38*, 299–308. [CrossRef]
4. Cassity-Duffey, K.; Cabrera, M.; Rema, J. Ammonia volatilization from broiler litter: Effect of soil water content and humidity. *Soil Sci. Soc. Am. J.* **2015**, *79*, 543–550. [CrossRef]
5. Lynch, D.; Henihan, A.M.; Bowen, B.; Lynch, D.; McDonnell, K.; Kwapinski, W.; Leahy, J.J. Utilisation of poultry litter as an energy feedstock. *Biomass Bioenerg.* **2013**, *49*, 197–204. [CrossRef]
6. Karunanithi, R.; Szogi, A.A.; Bolan, N.; Naidu, R.; Loganathan, P.; Hunt, P.G.; Vanotti, M.B.; Saint, C.P.; Ok, Y.S.; Krishnamoorthy, S. Phosphorus recovery and reuse from waste streams. *Adv. Agron.* **2015**, *131*, 173–250.

7. Jurgilevich, A.; Birge, T.; Kentala-Lehtonen, J.; Korhonen-Kurki, K.; Pietikäinen, J.; Saikku, L.; Schösler, H. Transition towards circular economy in the food system. *Sustainability* **2016**, *8*, 69. [CrossRef]
8. Tan, Z.; Lagerkvist, A. Phosphorus recovery from the biomass ash: a review. *Renew. Sust. Energ. Rev.* **2011**, *15*, 3588–3602. [CrossRef]
9. Chastain, J.P.; Coloma-del Valle, A.; Moore, K.P. Using broiler litter as an energy source: Energy content and ash composition. *Appl. Eng. Agric.* **2012**, *28*, 513–522. [CrossRef]
10. Codling, E.E.; Chaney, R.L.; Sherwell, J. Poultry litter ash as a potential phosphorus source for agricultural crops. *J. Environ. Qual.* **2002**, *31*, 954–961. [CrossRef] [PubMed]
11. Pagliari, P.H.; Rosen, C.J.; Strock, J.S. Turkey manure ash effects on alfalfa yield, tissue elemental composition, and chemical soil properties. *Comm. Soil Sci. Plant Anal.* **2009**, *40*, 2874–2897. [CrossRef]
12. Pagliari, P.H.; Rosen, C.J.; Strock, J.S. Characterization of turkey manure ash and its nutrient value for corn and soybean production. *Crop Manag.* **2009**. [CrossRef]
13. Szogi, A.A.; Vanotti, M.B. Prospects of phosphorus recovery from poultry litter. *Bioresour. Technol.* **2008**, *100*, 5461–5465. [CrossRef]
14. Demeyer, A.; Voundi Nkana, J.C.; Verloo, M.G. Characteristics of wood ash and influence on soil properties and nutrient uptake: An overview. *Bioresour. Technol.* **2001**, *77*, 287–295. [CrossRef]
15. Codling, E.E. Laboratory characterization of extractable phosphorus in poultry litter and poultry litter ash. *Soil Sci.* **2006**, *171*, 858–864. [CrossRef]
16. Bogush, A.A.; Stegemann, J.A.; Williams, R.; Wood, I.G. Element speciation in UK biomass power plant residues based on composition, mineralogy, microstructure and leaching. *Fuel* **2018**, *211*, 712–725. [CrossRef]
17. Pagani, A.; Mallarino, A.P. Soil pH Change over Time as Affected by Sources and Application Rates of Liming Materials. Iowa State Research Farm Progress Reports 259. 2011. Available online: https://lib.dr.iastate.edu/cgi/viewcontent.cgi?article=1269&context=farms_reports (accessed on 16 March 2019).
18. Chen, L.; Dick, W.A. *Gypsum as an Agricultural Amendment: General Use Guidelines*; The Ohio State University Extension: Columbus, OH, USA, 2011.
19. American Coal Ash Association. 2016 Coal Combustion Product (CCP) Production and Use Survey Report. Available online: https://www.acaa-usa.org/Portals/9/Files/PDFs/2016-Survey-Results.pdf (accessed on 16 March 2019).
20. Codling, E.E.; Lewis, J.; Watts, D.B. Broiler litter ash and flue gas desulfurization gypsum effects on peanut yield and uptake of nutrients. *Comm. Soil Sci. Plant Anal.* **2015**, *46*, 2553–2575. [CrossRef]
21. Unit III 5.5: Digestion and dissolution methods for P, K, Ca, Mg, and trace elements. In *Recommended Methods of Manure Analysis (A3769)*; Peters, J. (Ed.) University of Wisconsin-Extension Publication: Madison, WI, USA, 2003.
22. AOAC International. *Official Methods of Analysis*, 17th ed.; AOAC: Gaithersburg, MD, USA, 2000.
23. Sikora, F.J.; Moore, K.P. *Soil Test Methods of from the Southeastern United States, Southern Cooperative Series Bulletin No. 419*; Southern Extension and Research Activity-Information Exchange Group 6: Clemson, SC, USA, 2014.
24. ASABE Standards S341.5. *Procedure for Measuring Distribution Uniformity and Calibrating Granular Broadcast Spreaders*; American Society of Agricultural and Biological Engineers: St. Joseph, MI, USA, 2018.
25. Obernberger, I.; Brunner, T.; Barnthaler, G. Chemical properties of solid biofuels-significance and impact. *Biomass Bioenerg.* **2006**, *30*, 973–982. [CrossRef]
26. Novak, J.M.; Szogi, A.A.; Watts, D.W. Copper and zinc accumulation in sandy soils and constructed wetlands receiving pig manure effluent application. In *Trace Elements in Animal Production Systems*; Schlegel, P., Durosoy, S., Jongbloed, A.W., Eds.; Wageningen Academic Publishers: Wageningen, The Netherlands, 2008; pp. 45–54.
27. Clarholm, M. Granulated wood ash and a 'N-free' fertilizer to a forest soil-effects on P availability. *Forest Ecol. Manag.* **1994**, *66*, 127–136. [CrossRef]
28. Moore, P.A., Jr.; Miller, D.M. Decreasing phosphorus solubility in poultry litter with aluminum, calcium and iron amendments. *J. Environ. Qual.* **1994**, *23*, 325–330. [CrossRef]
29. Anderson, D.L.; Tuovinen, O.H.; Faber, A.; Ostrokowski, I. Use of soil amendments to reduce soluble phosphorus in dairy soils. *Ecol. Eng.* **1995**, *5*, 229–246. [CrossRef]
30. Dou, Z.; Zhang, G.Y.; Stout, W.L.; Toth, J.D.; Ferguson, J.D. Efficacy of alum and coal combustion by-products in stabilizing manure phosphorus. *J. Environ. Qual.* **2003**, *32*, 1490–1497. [CrossRef] [PubMed]

31. Brauer, D.; Aiken, G.E.; Pote, D.H.; Livingston, S.J.; Norton, L.D.; Way, T.R.; Edwards, J.H. Amendments effects on soil test phosphorus. *J. Environ. Qual.* **2005**, *34*, 1682–1686. [CrossRef] [PubMed]
32. Watts, D.B.; Torbert, H.A. Impact of gypsum applied to grass buffer strips on reducing soluble P in surface water runoff. *J. Environ. Qual.* **2009**, *38*, 1511–1517. [CrossRef] [PubMed]
33. Endale, D.M.; Schomberg, H.H.; Fisher, D.S.; Franklin, D.H.; Jenkins, M.B. Flue gas desulfurization gypsum: implication for runoff and nutrient losses associated with broiler litter use on pastures on Ultisols. *J. Environ. Qual.* **2014**, *43*, 281–289. [CrossRef]
34. Adotey, N.; Harrell, K.L.; Weatherford, W.P. Characterization and liming effect of wood ash generated from a biomass-fueled commercial power plant. *Comm. Soil Sci. Plant Anal.* **2018**, *49*, 38–49. [CrossRef]
35. Havlin, J.L.; Tisdale, S.L.; Nelson, W.L.; Beaton, J.D. *Soil Fertility and Fertilizers: An Introduction to Nutrient Management*, 6th ed.; Prentice Hall: Upper Saddle River, NJ, USA, 1999.
36. Pagliari, P.; Rosen, C.; Strock, J.; Russelle, M. Phosphorus availability and early corn growth response in soil amended with turkey manure ash. *Comm. Soil Sci. Plant Anal.* **2010**, *41*, 1369–1382. [CrossRef]
37. Wells, D.E.; Beasley, J.S.; Bush, E.W.; Gaston, L.A. Poultry litter ash rate and placement affect phosphorus dissolution in a horticultural substrate. *J. Environ. Hort.* **2017**, *35*, 117–127.
38. Wells, D.E.; Beasley, J.S.; Gaston, L.A.; Bush, E.W.; Thiessen, M.E. Poultry litter ash reduces phosphorus losses during greenhouse production of *Lantana camara* L. 'New Gold'. *HortScience* **2017**, *52*, 592–597. [CrossRef]
39. Sumner, P.E. *Calibration of Bulk Dry Fertilizer Applicators, The University of Georgia Cooperative Extension Circular 798*; University of Georgia College of Agricultural Environmental Sciences: Athens, GA, USA, 2012.
40. Smith, D.B.; Willcutt, M.H.; Doler, J.C.; Diallo, Y. Uniformity of granular fertilizer applications with a spinner truck. *App. Eng. Agric.* **2004**, *20*, 289–295. [CrossRef]
41. Kemp, T.; Carolina Eastern Inc., Charleston, SC, USA. Personal communication, 2018.

environments

Article

Phytostabilization of Zn and Cd in Mine Soil Using Corn in Combination with Biochars and Manure-Based Compost

Gilbert C. Sigua [1,*], Jeff M. Novak [1], Don W. Watts [1], Jim A. Ippolito [2], Thomas F. Ducey [1], Mark G. Johnson [3] and Kurt A. Spokas [4]

1 Department of Agriculture, Agricultural Research Service, Coastal Plains Soil, Water, and Plant Research Center, Florence, CA 29501, USA; jeff.novak@ars.usda.gov (J.M.N.); don.watts@ars.usda.gov (D.W.W.); tom.ducey@ars.usda.gov (T.F.D.)

2 Department of Soil and Crop Sciences, C006 Plant Sciences Building, Colorado State University, Fort Collins, CO 80523, USA; jim.ippolito@colostate.edu

3 United States Environmental Protection Agency, National Health and Environmental Effects Research Laboratory, Western Ecology Division, 200 Southwest 35th Street, Corvallis, OR 97333, USA; johnson.markg@epa.gov

4 Department of Agriculture, Agricultural Research Service, St. Paul, MN 55108, USA; kurt.spokas@ars.usda.gov

* Correspondence: gilbert.sigua@ars.usda.gov; Tel.: +1-843-669-5203

Received: 30 April 2019; Accepted: 5 June 2019; Published: 13 June 2019

Abstract: Mining activities could produce a large volume of spoils, waste rocks, and tailings, which are usually deposited at the surface and become a source of metal pollution. Phytostabilization of the mine spoils could limit the spread of these heavy metals. Phytostabilization can be enhanced by using soil amendments such as manure-based biochars capable of immobilizing metal(loid)s when combined with plant species that are tolerant of high levels of contaminants while simultaneously improving properties of mine soils. However, the use of manure-based biochars and other organic amendments for mine spoil remediation are still unclear. In this greenhouse study, we evaluated the interactive effect of biochar additions (BA) with or without the manure-based compost (MBC) on the shoots biomass (SBY), roots biomass (RBY), uptake, and bioconcentration factor (BCF) of Zn and Cd in corn (*Zea mays* L.) grown in mine soil. Biochar additions consisting of beef cattle manure (BCM); poultry litter (PL); and lodge pole pine (LPP) were applied at 0, 2.5, and 5.0% (w/w) in combination with different rates (0, 2.5, and 5.0%, w/w) of MBC, respectively. Shoots and roots uptake of Cd and Zn were significantly affected by BA, MBC, and the interaction of BA and MBC. Corn plants that received 2.5% PL and 2.5% BCM had the greatest Cd and Zn shoot uptake, respectively. Corn plants with 5% BCM had the greatest Cd and Zn root uptake. When averaged across BA, the greatest BCF for Cd in the shoot of 92.3 was from the application of BCM and the least BCF was from the application of PL (72.8). Our results suggest that the incorporation of biochar enhanced phytostabilization of Cd and Zn with concentrations of water-soluble Cd and Zn lowest in soils amended with manure-based biochars while improving the biomass productivity of corn. Overall, the phytostabilization technique and biochar additions have the potential to be combined in the remediation of heavy metals polluted soils.

Keywords: biochar; phytoextraction; corn; uptake; mine soils; heavy metals; root biomass; shoot biomass

1. Introduction

Mining activities usually produce a large volume of spoils, waste rocks, and tailings, which are usually deposited at the soil surface. If the spoils contain heavy metals that are soluble, there is a potential of heavy metal pollution contamination and off-site movement. Mined areas near Webb City in Jasper County, Missouri, contained mine waste piles that were removed, but still provide a source of heavy metal contamination, particularly Zn and Cd in the underlying soil. Mining activities can lead to extensive environmental pollution of terrestrial ecosystem due to the deposition of heavy-metal containing waste materials, tailings, and lagoon wastes [1–3].

Metal (loid) contaminants such as Cd and Zn are significant issues, not only for the environment, but especially for human health [4–6]. These contaminated areas present a health risk and are recognized as areas that need to be remediated to allow for crop phytostabilization to occur [1]. Often, contaminated sites are not conducive for plant growth due to metal toxicity, lack of soil nutrients, low pH values, poor microbial activity, and unsuitable physical soil properties. Both physical and chemical techniques have been considered in mine spoil remediation, but these methods have flaws, are expensive, and can be disruptive to soils. Remediation of these contaminated and hazardous soils by conventional practices using excavation and landfilling is arguably unfeasible on large scales because these techniques are cost-prohibitive and environmentally disruptive [7,8]. Phytostabilization techniques that involve the establishment of plant cover on the surface of contaminated sites could serve as an efficient alternative remediation approach as they provide low-cost and environmentally friendly options [7,9]. For this reason, remediation of contaminated sites using phytostabilization techniques require the amendment to improve soil-plant relationships thereby stimulating plant growth.

Remediation of mine spoil can be a complex process due to several chemical and physical factors that can limit plant growth [10]. Bolan et al. [11] summarized the different factors affecting phytostabilization. For example, soil, plant, contaminants, and environmental factors determine the successful outcome of phytostabilization technology in relation to both the remediation and revegetation of contaminated sites. Mine spoils may have unfavorable soil chemical characteristics, e.g., very low pH, phytotoxic metals [12,13], physical limitations (e.g., high bulk density, low soil moisture retention, poor aggregation [14]; and unsuitable microbial habitat conditions, e.g., low soil organic matter and poor nutrient turnover [15]. These aspects can severely limit plant growth. As such, reclamation plans usually involve applying soil amendments (i.e., composts, lime) to neutralize their low pH, and to raise organic matter levels that favors organic binding of metals, along with enhanced microbial enzymatic activity for nutrient cycling [16].

Phytostabilization can be enhanced by using soil amendments that immobilize metal(loid)s when combined with plant species that are tolerant of high levels of contaminants while simultaneously improving the physical, chemical, and biological properties of mine soils. Some previously used amendments to improve soil conditions include biosolids, lime, green waste, or biochars. Among these amendment types, the use of biochar has recently been investigated for in situ remediation of contaminated lands in association with plants [10,17–19]. The incorporation of organic amendments improves the quality of mine soils and makes it possible for vegetation to be established [20,21]. Recent studies have highlighted that biochars are effective soil amendments in that they improve soil conditions to raise the agronomic values of soils [22–25].

Numerous studies may have shown that adding organic amendments (e.g., biochars, sewage sludge, manures) to soil promotes the phytoextraction process [26,27], but only few studies have evaluated the combined effect of organic amendments and phytostabilization with corn in Cd and Zn contaminated mine soils. There is a lack of agreement over the influence of organic amendments such as biochars on metal immobilization in soil. Moreover, the application of biochars to contaminated soil systems has not been systematically investigated to any great extent. Biochar may be a tool for mine spoil remediation; however, its mechanisms for achieving this goal are still not well understood. The objective of our study was to evaluate the interactive effects of biochar additions with or without the manure-based compost (MBC) on shoots biomass (SBY), roots biomass (RBY), uptake, and bioconcentration factor (BCF) of Zn and Cd in corn (*Zea mays* L.) grown in mine soil.

2. Materials and Methods

2.1. Site Description, Soil Characterization, and Soil Preparation

A field for sampling soil was selected near Webb City in Jasper County, MO (latitude 37.13°, longitude 94.45°). This location is a part of the Oronogo-Duenweg mining area of Southwest MO. Mining of lead (Pb) and zinc (Zn) ore has occurred across the country with leftover milling waste discarded in chat piles. The chat piles contain residual Pb and Zn concentrations that in some locations moved into the underlying soil.

Prior to the mining disturbance, soil in this field was mapped as a Rueter series, which is classified using United States Department of Agriculture Taxonomic terminology as a loamy-skeletal, siliceous, active, mesic Typic Paleudalf. Examination of the Reuter soil profile reveals that it has extremely gravelly silt loam textured soil horizons that formed in colluvium over residuum derived from limestone (Soil Survey of Jasper County, MO, 2002).

For our purposes, a backhoe was used to collect a few hundred kg of C horizon material down from 60- to 90-cm deep. The soil along with coarse fragments was placed in plastic-lined metal drums and transported to the ARS-Florence (Florence, SC, USA). The C horizon material was removed from the drums and air-dried. As a result of the presence of large cobbles, the soil was screened using a 12.7-mm diameter sieve to collect soil material more appropriate for use in a potted greenhouse experiment. Sieving the soil revealed that it contained approximately 30% (w/w) coarse fragments that were >12.7-mm in diameter. Soil that passed through the sieve was stored in the plastic line drums for characterization and used in our greenhouse experiment.

The sieve C horizon material (<12.7-cm diameter) was characterized for its pH (4.40) using a 1:2 (w/w) soil:deionized water ratio [16]. Additionally, bioavailable metal and total metal concentrations were extracted using multiple extractants and acid digestion, respectively. Both deionized water (water-soluble) and 0.01M $CaCl_2$ (extractable) metal concentrations were determined in triplicate by extracting 30 g soil with 60 mL of liquid extractant, shaken for 30 m, and filtered using a nylon 0.45 μM filter syringe [10,16,28]. Extraction with diethylenetriamine pentaacetic acid (DTPA) was also conducted in triplicate using 10 g of soil with 20 mL of DTPA after shaking for 2 h, and filtration using 0.45 μm filter syringe [10,16]. Total metal concentrations were determined in triplicate by digestion of 10 g soil in 100 mL of 4 M HNO_3 as described [28]. All water-soluble and extractable metal concentrations including Cd and Zn were quantified via the inductively coupled plasma spectroscopy atomic emission spectroscopy (ICP-AES) (Thermo Fisher Scientific, West Palm Beach, FL, USA). Concentrations of Cd and Zn and other chemical properties of C horizon are presented in Table 1.

Table 1. Chemical properties of Tri-State Mine soil (C horizon) used in the study.

		Extractable Metal (mg kg^{-1})	
Element	**Total Metal (mg kg^{-1}) †**	**H_2O**	**0.01 M $CaCl_2$**
Al	ND	4.40 ± 3.87	11.36 (1.56)
Cl	ND	ND	ND
Cd	72.2 (2.7)	5.73 ± 0.98	50.45 (0.40)
Cr	ND	0	0.12 (0.01)
Cu	66.5 (2.5)	0.22 (0.07)	2.17 (0.02)
Fe	ND	10.57 (2.11)	12.65 (1.45)
K	711 (25)	26.18 (3.47)	59.48 (3.35)
Mg	355 (45)	4.53 (1.38)	36.49 (0.97)
Mn	72 (5.7)	2.48 (0.72)	21 (0.9)
Na	ND	22.25 (4.01)	25.58 (4.24)
Ni	7.6 (0.3)	0.18 (0.01)	0.45 (0.01)
P	168 (4)	3.89 (0.06)	1.43 (1.30)
Pb	23.5 (0.7)	0	0
SO_4	ND	152.6 (19.4)	112.8 (17.4)
Zn	2225 (12)	141.0 (25.7)	782 (13)

† samples digested using 4M HNO_3; (means of n = 3; standard deviation in parentheses; ND = not determined; 0 value = below detection limit).

2.2. Experimental Setup and Design

The experimental treatments consisted of biochar additions (BA): Beef cattle manure (BCM); poultry litter (PL); and lodge pole pine (LPP) that were applied at 0, 2.5, and 5.0% (w/w) in combination with different rates (0, 2.5, and 5.0%, w/w) of MBC (RMBC), respectively. Experimental treatments were replicated three times using a 3 × 2 × 3 split plot arrangement in completely randomized block design.

The treated and untreated C material soils were placed into triplicate plastic flower pots (15-cm top diameter × 17-cm deep) and gently tapped to a bulk density of 1.5 g/cm^3 as outlined in Novak et al. [16]. Eight corn seeds were then planted in each pot. The pots were transported to a greenhouse and randomly placed on benches. Corn in the pots were kept in the greenhouse under a mean air temperature of about 21.8 ± 3.1 °C and relative humidity of about 53 ± 12.2%. On day 16, all pots were fertilized with a 10 mL solution of NH_4NO_3 that delivered an equivalent of 25 kg N ha-1 because some treatments exhibited N deficient response in corn leaves (yellowing). No inorganic P or K was added to the pots because these nutrients were supplied with the amendments. The pots were watered by hand using recycled water several times per week.

2.3. Feedstock Collection, Description, Biochar Production, and Characterization

Three feedstocks were used to produce biochars in this experiment namely: Beef cattle manure; lodge pole pine; and poultry litter. The raw beef cattle manure was collected from a local feedlot operation near Webb City, MO. The manure pile was exposed to the environment for 1–2 years to allow for conversion into a manure-based compost mixture. A few kg of the manure compost was transported to the ARS-Florence location and sieved using a 6-mm sieve. A portion of the 6-mm sieved manure compost was pyrolyzed at 500 °C into biochar as outlined in Novak et al. [29]. The remaining two biochars were available commercially and consisted of biochar produced from the poultry litter and lodgepole pine feedstocks. The poultry litter biochar was produced by gasification using a fixed-bed pyrolyzer and the lodgepole pine biochar was produced using a slow pyrolysis process. The pyrolysis temperatures employed to produce these two biochars are not available.

All three biochars were characterized for their pH and electrical conductivity in a 1:2 (w/w) biochar to deionized water ratio [16]. All three biochars were also characterized chemically (ASTM D3176; Hazen Research, Inc., Golden, CO, USA). The molar H/C and O/C ratios were calculated from the elemental analysis. Total elemental composition of all three biochars was determined using concentrated HNO_3 acid digestion described in the US EPA 305b method [29,30] and were quantified using an inductively coupled plasma atomic emission spectroscopy (ICP-AES). Similar characterization was performed on the beef cattle manure compost feedstock as described above. Some of the chemical and physical properties of the manure-based compost and biochars are shown in Table 2. The appropriateness of using the different designer biochars in our study were based on an early published paper by Novak et al. [16].

Table 2. Chemical and physical properties of compost and biochars (dry-basis).

A. Ultimate and Proximate Analysis				
	Beef Cattle Manure		Lodgepole Pine	Poultry Litter
Measurement (%)	Compost	Biochar	Biochar	Biochar
C	17.5	13.8	90.5	37.4
H	1.9	0.7	2.4	2.8
O	10.5	1.4	3.2	13.0
N	1.6	1.0	0.7	4.2
S	0.09	0.02	<0.001	007
Ash	68.4	83.1	3.2	42.5
Fixed C	6.1	9.4	82.5	21.2
Volatile matter	25.5	7.5	14.3	36.3
pH	6.8	9.5	9.7	9.1
O/C	0.46	0.07	0.03	0.26
H/C	1.29	0.60	0.32	0.89
B. Elemental Analysis of Ash (%, Ash wt Basis)				
Al	3.0	2.9	0.9	0.9
As	<0.005	<0.005	0.1	<0.005
Ca	3.0	2.8	11.8	11.6
Cd	<0.005	<0.005	<0.005	<0.005
Cl	<0.01	<0.01	0.6	5.6
Cr	<0.005	<0.005	0.15	0.01
Cu	0.005	0.005	0.26	0.4
Fe	1.43	1.41	1.13	1.11
K	2.2	2.13	3.9	18.0
Mg	0.93	0.90	2.6	3.9
Mn	0.09	0.10	0.35	0.28
Na	0.31	0.30	1.1	4.5
Ni	0.005	0.006	0.03	0.016
P	0.67	0.68	0.4	8.6
Pb	<0.005	<0.005	0.09	<0.005
S	0.25	0.22	0.58	4.9
Si	77.6	77.2	18.2	8.4
Zn	0.03	0.03	0.09	0.23

2.4. *Tissue Analyses for Cadmium and Zinc Concentrations in Shoots and Roots of Corn*

At day 35, corn roots were observed to grow out of the pot bottoms. The experiment was terminated, and the corn shoots and roots were harvested from each pot, oven-dried (60 °C), and digested as described by Hunag and Schulte [31]. Snipped samples were digested in an auto-block using a mixture of nitric and hydrogen peroxide. The concentrations of Cd and Zn in the tissues were analyzed using an ICP spectroscopy. Tissue uptake of Cd and Zn were calculated using Equation (1) for the shoot's uptake and Equation (2) for the root's uptake.

$$MU_{Cd,\,Zn} = [MBC_{d,\,Zn}] \times SBY \quad (1)$$

where: MU = metal uptake (kg ha^{-1}); CM = concentration of Cd and Zn (%) in corn shoot tissues; SBY = dry matter yield of shoots (kg ha^{-1}).

$$MU_{Cd,\,Zn} = [MBC_{d,\,Zn}] \times RBY \quad (2)$$

where: MU = metal uptake (kg ha^{-1}); CM = concentration of Cd and Zn (%) in corn root tissues; RBY = dry matter yield of roots (kg ha^{-1}).

2.5. *Bioconcentration Factor of Cd and Zn in Shoots and Roots of Corn*

The bioconcentration factor (BCF) in corn was calculated as the ratio between heavy metal concentration in the plants (shoots and roots) and the total heavy metal in the soil as shown in Equations (3) and (4).

$$BCF_{shoots} = [MBC_{d,\,Zn}]_{shoots}/[MBC_{d,\,Zn}]_{soils} \quad (3)$$

$$BCF_{roots} = [MBC_{d,\,Zn}]_{roots}/[MBC_{d,\,Zn}]_{soils} \quad (4)$$

where: BCF_{roots} = bioconcentration factor for Cd and Zn in the roots of corn; BCF_{shoots} = bioconcentration factor for Cd and Zn in the shoots of corn; CM_{shoot} = concentration of Cd and Zn (%) in the corn shoot; and CM_{soils} = concentration of Cd and Zn (%) in the soil.

2.6. *Statistical Analysis*

To determine the effect of different biochar additions (BA) and rates of biochar additions (BR) with or without the manure-based compost (MBC) on biomass and uptake (Cd and Zn) of corn grown in mine soils, data were analyzed with a three-way ANOVA using PROC GLM [32]. For this study, the *F*-test indicated significant results at 5% level of significance, so means of the main treatments (additions of biochars, BA), sub-treatments (rates of biochar additions, BR), sub-sub treatments (rates of MBC, RMBC) were separated following the procedures of the least significance differences (LSD) test, using appropriate mean squares [32].

3. Results

3.1. *Soil pH and Water-Soluble Cd and Zn Concentrations in Mine Soils*

Soil pH and concentrations of water-soluble Cd and Zn in mine spoil soils varied significantly with BA ($p \leq 0.0001$), BR ($p \leq 0.0001$), and RMBC ($p \leq 0.0001$). While soil pH was not affected by the interaction effect of BR × RMBC, soil pH and concentrations of Cd and Zn in the soils were significantly affected by the interactions of BA × BR × RMBC (Table 3). Incorporation of 5% PL with 5% RMBC resulted in significantly higher soil pH (6.61 ± 0.01), but significantly lower concentrations of Cd (0.63 ± 0.16 mg kg^{-1}) and Zn (10.69 ± 1.95 mg kg^{-1}) when compared with the control soils (pH of 4.73 ± 0.32; Cd of 1.89 ± 0.35 mg kg^{-1}; Zn of 63.89 ± 11.08 mg kg^{-1}).

Table 3. Average concentrations of water-soluble Cd and Zn and pH in mine spoil soil.

Biochar Additions	Biochar Rate (%)	Compost Rate (%)	pH	Cd (mg/kg)	Zn (mg/kg)
Control	0	0	4.40 ± 0.06	2.05 ± 0.22	62.06 ± 6.21
		2.5	4.69 ± 0.05	2.12 ± 0.13	70.38 ± 4.20
		5.0	5.10 ± 0.03	1.51 ± 0.08	57.12 ± 9.68
	Mean		**4.73 ± 0.32**	**1.89 ± 0.35**	**63.89 ± 11.08**
Beef Cattle Manure	2.5	0	5.07 ± 0.14	1.75 ± 0.15	56.32 ± 5.06
		2.5	5.19 ± 0.07	1.37 ± 0.11	51.11 ± 3.51
		5.0	5.28 ± 0.12	1.10 ± 0.05	42.77 ± 2.72
	Mean		**5.18 ± 0.13**	**1.41 ± 0.29**	**49.73 ± 7.22**
	5.0	0	5.31 ± 0.22	1.68 ± 0.14	53.81 ± 3.81
		2.5	5.61 ± 0.14	1.04 ± 0.15	37.25 ± 4.52
		5.0	5.91 ± 0.14	0.94 ± 0.26	32.85 ± 7.84
	Mean		**5.61 ± 0.30**	**1.22 ± 0.39**	**41.31 ± 10.76**
Lodge Pole Pine	2.5	0	4.37 ± 0.01	2.57 ± 0.59	75.22 ± 7.26
		2.5	4.77 ± 0.07	2.31 ± 0.12	75.08 ± 4.69
		5.0	5.10 ± 0.03	1.50 ± 0.04	53.27 ± 1.10
	Mean		**4.75 ± 0.26**	**2.13 ± 0.57**	**67.85 ± 6.14**
	5.0	0	4.47 ± 0.02	2.56 ± 0.04	70.86 ± 1.96
		2.5	4.89 ± 0.10	1.69 ± 0.32	52.35 ± 9.91
		5.0	5.05 ± 0.05	2.04 ± 0.27	68.47 ± 9.21
	Mean		**4.81 ± 0.26**	**2.08 ± 0.44**	**63.89 ± 11.08**
Poultry Litter	2.5	0	5.46 ± 0.16	3.38 ± 0.89	94.02 ± 22.62
		2.5	5.58 ± 0.24	1.94 ± 0.02	60.48 ± 6.42
		5.0	5.85 ± 0.02	1.49 ± 0.13	47.53 ± 3.42
	Mean		**5.63 ± 0.23**	**2.27 ± 0.98**	**67.35 ± 23.93**
	5.0	0	6.33 ± 0.03	1.19 ± 0.02	20.57 ± 1.17
		2.5	6.53 ± 0.01	0.84 ± 0.07	13.28 ± 1.08
		5.0	6.61 ± 0.01	0.63 ± 0.16	10.69 ± 1.95
	Mean		**6.49 ± 0.13**	**0.89 ± 0.26**	**14.85 ± 4.61**
Sources of Variation				**Level of Significance**	
Biochar Additions (BA)			***	***	***
Biochar Rate (BR)			***	***	***
Compost Rate (RMBC)			ns	***	***
BA × BR			***	***	***
BA × RMBC			**	**	***
BR × RMBC			ns	ns	ns
BA × BR × RMBC			ns	**	*

*** Significant at $p \leq 0.0001$; ** Significant at $p \leq 0.001$; * Significant at $p \leq 0.01$; ns – not significant.

Of the different additions of biochar (BA) when averaged across BR and RMBC, the greatest soil pH increase was from soil treated with PL (6.06 ± 0.18) followed by BCM (5.39 ± 0.21), LPP (4.78 ± 0.26) and control soil (4.73 ± 0.32). The effect of BA on water-soluble Cd (mg kg^{-1}) is as follows: LPP (2.10 ± 0.51) > control (1.89 ± 0.35) > PL (1.58 ± 0.62) > BCM (1.32 ± 0.34). The greatest average concentration of water-soluble Zn (mg kg^{-1}) was from soil treated with LPP (65.87 ± 8.61) followed by control soil (63.89 ± 11.08), BCM (45.52 ± 8.99), and PL (41.10 ± 28.54) (Table 3).

Overall, the pH of mine soils was significantly affected by the increasing rate (2.5% to 5.0%) of different BA (Table 2). The soil pH of mine soils treated with 2.5% and 5.0% BCM was increased from 5.18 ± 0.13 to 5.61 ± 0.30. Similarly, the pH of soils treated with 2% and 5% LPP was increased from 4.75 ± 0.26 to 4.81 ± 0.26. A much higher increase in the pH of mine soils when treated with 2.5% PL (5.63 ± 0.23) and 5% PL (6.49 ± 0.13). On the other hand, the concentration of water-soluble Cd showed a decreasing trend with the increasing rate of BA application (i.e., 2.5% to 5%). The concentration of water-soluble Cd (mg kg^{-1}) in soils was reduced from 1.41 ± 0.29 to 1.22 ± 0.39; 2.13 ± 0.57 to 2.08 ± 0.44; and 2.27 ± 0.89 to 0.89 ± 0.26 when treated with 2.5% and 5% BCM; LPP; and PL, respectively. The concentrations of Cd in the soils were also reduced significantly following the addition of raw beef cattle manure (Table 3). The concentrations of water-soluble Zn (mg kg^{-1}) in the soil also showed decreasing trends following the additions of increasing rates of biochars and beef cattle manure compost. The concentration of water-soluble Zn (mg kg^{-1}) in soils was reduced from 49.73 ± 7.22 to 41.31 ± 10.76; 67.85 ± 6.14 to 63.89 ± 11.08; and 67.35 ± 23.93 to 14.85 ± 4.61 when treated with 2.5% and 5% BCM; LPP; and PL, respectively. Again, results have shown the beneficial effects of increasing rates of biochar in combination with the increasing rates application of compost beef cattle manure on enhancing the soil pH while decreasing the concentrations of water-soluble Cd and Zn in mine soils.

3.2. *Concentrations of Cd and Zn in Corn Shoots and Roots*

Except for the concentration of Cd in the shoots, all other concentrations of Cd and Zn in the shoots and roots varied significantly with BA ($p \leq 0.0001$), BR ($p \leq 0.0001$), and RMBC ($p \leq 0.0001$). The interactions of BA × BR and BA × RMBC showed highly significant effects on the Cd and Zn concentrations both in corn shoots and roots (Table 4).

Table 4. Average concentrations of Cd and Zn in shoots and roots biomass of corn.

Biochar Additions	Biochar Rate (%)	Compost Rate (%)	Cd (mg/kg)	Zn (mg/kg)	Cd (mg/kg)	Zn (mg/kg)
			Shoots		Roots	
Control	0	0	210.7 ± 49.8	3485.3 ± 874.6	150.1 ± 29.2	3235.2 ± 354.4
		2.5	145.5 ± 20.9	3870.1 ± 512.4	255.3 ± 67.2	3686.7 ± 801.8
		5.0	99.1 ± 12.8	3165.5 ± 363.6	246.9 ± 19.5	3531.7 ± 240.2
	Mean		**151.7 ± 55.9**	**3506.9 ± 477.3**	**217.5 ± 62.7**	**3484.5 ± 496.0**
Beef Cattle Manure	2.5	0	202.8 ± 20.9	4881.1 ± 239.3	270.9 ± 32.7	4390.2 ± 442.9
		2.5	123.1 ± 17.3	3591.3 ± 313.5	277.2 ± 31.9	3569.1 ± 466.1
		5.0	96.7 ± 7.1	2716.7 ± 151.6	245.7 ± 50.4	2863.8 ± 211.5
	Mean		**140.8 ± 49.9**	**3729.7 ± 966.3**	**264.6 ± 36.9**	**3607.7 ± 512.8**
	5.0	0	178.5 ± 7.4	4437.5 ± 42.9	282.3 ± 44.0	3723.2 ± 266.3
		2.5	99.2 ± 8.3	2508.5 ± 282.6	216.8 ± 18.8	2681.2 ± 158.9
		5.0	69.1 ± 0.4	1575.2 ± 121.6	188.7 ± 45.3	2053.1 ± 417.6
	Mean		**115.6 ± 49.3**	**2840.4 ± 273.7**	**229.3 ± 42.8**	**2819.2 ± 512.8**
Lodge Pole Pine	2.5	0	154.4 ± 59.9	2611.1 ± 123.9	151.2 ± 38.7	2666.9 ± 557.3
		2.5	170.4 ± 26.9	4145.5 ± 448.9	228.1 ± 74.3	3273.6 ± 736.1
		5.0	155.6 ± 16.9	4236.9 ± 618.1	229.2 ± 3.0	3102.9 ± 194.2
	Mean		**160.1 ± 34.8**	**3664.5 ± 440.4**	**202.9 ± 57.1**	**3014.5 ± 554.0**
	5.0	0	214.3 ± 42.8	3273.8 ± 645.9	152.9 ± 16.9	2933.0 ± 498.4
		2.5	167.1 ± 23.2	3920.8 ± 340.7	172.2 ± 38.1	2985.4 ± 432.2
		5.0	139.8 ± 12.1	3577.4 ± 252.6	210.3 ± 36.1	2850.4 ± 253.9
	Mean		**173.7 ± 41.2**	**3590.7 ± 477.3**	**178.5 ± 37.4**	**2922.9 ± 358.3**
Poultry Litter	2.5	0	231.4 ± 21.2	3127.1 ± 112.9	227.9 ± 45.2	2222.9 ± 177.9
		2.5	160.6 ± 13.1	2227.8 ± 171.4	256.8 ± 77.6	2101.7 ± 170.4
		5.0	126.2 ± 11.6	1681.3 ± 157.2	159.8 ± 23.7	1892.5 ± 287.8
	Mean		**172.7 ± 48.1**	**2345.4 ± 158.9**	**214.9 ± 63.4**	**2072.3 ± 238.4**
	5.0	0	79.3 ± 17.4	651.8 ± 130.5	87.8 ± 15.5	982.9 ± 158.9
		2.5	55.4 ± 10.6	467.2 ± 72.5	51.6 ± 5.4	623.3 ± 125.4
		5.0	50.72 ± 5.7	474.8 ± 65.7	53.2 ± 5.4	655.1 ± 114.1
	Mean		**61.9 ± 16.9**	**531.3 ± 121.8**	**64.2 ± 18.8**	**753.8 ± 116.8**
Sources of Variation			Level of Significance			
Biochar Additions (BA)			***	***	***	***
Biochar Rate (BR)			***	***	***	***
Compost Rate (RMBC)			ns	**	***	***
BA × BR			***	***	***	***
BA × RMBC			**	***	**	***
BR × RMBC			ns	ns	ns	ns
BA × BR × RMBC			ns	ns	**	*

*** Significant at $p \leq 0.0001$; ** Significant at $p \leq 0.001$; * Significant at $p \leq 0.01$; ns – not significant.

Overall, the concentrations of Cd and Zn in the shoots and roots with different additions of biochars when averaged across BR and RMBC were significantly lower than the concentrations of Cd and Zn in the shoots and roots of untreated corn. Applications of 2.5% and 5% PL resulted in the most significant reductions of Cd and Zn concentrations (mg kg^{-1}) in the shoots and roots of corn when compared with BCM and LPP with mean values of 172.7 ± 48.1 to 61.9 ± 16.9; 531.3 ± 121.8 to 214.9 ± 63.4; and 2354.4 ± 158.9 to 531.3 ± 121.8; and 2072.3 ± 238.4 to 753.8 ± 116.8, respectively (Table 4). These values were significantly lower than the concentrations of Cd and Zn both in the shoots and roots of untreated corn, suggesting the beneficial effects of biochar applications in phytostabilizing Cd and Zn using corn in mine soils.

3.3. Corn Shoots and Roots Biomass

The greatest total corn biomass (kg ha^{-1}) was from soils treated with PL (7122.3) followed by BCM (7005.6), and LPP (5008.7). The lowest total biomass of corn was from the untreated soils with a mean value of 5201.6 kg ha^{-1} (Figure 1). The shoot biomass varied significantly with BA ($p \leq 0.0001$) and RMBC ($p \leq 0.0001$), but not with BR (Table 5). On the other hand, the root biomass varied significantly with BA ($p \leq 0.0001$), BR ($p \leq 0.05$), and RMBC ($p \leq 0.05$). The interaction effects of BA × BR × RMBC failed to significantly affect the shoots and roots biomass of corn (Tables 5 and 6).

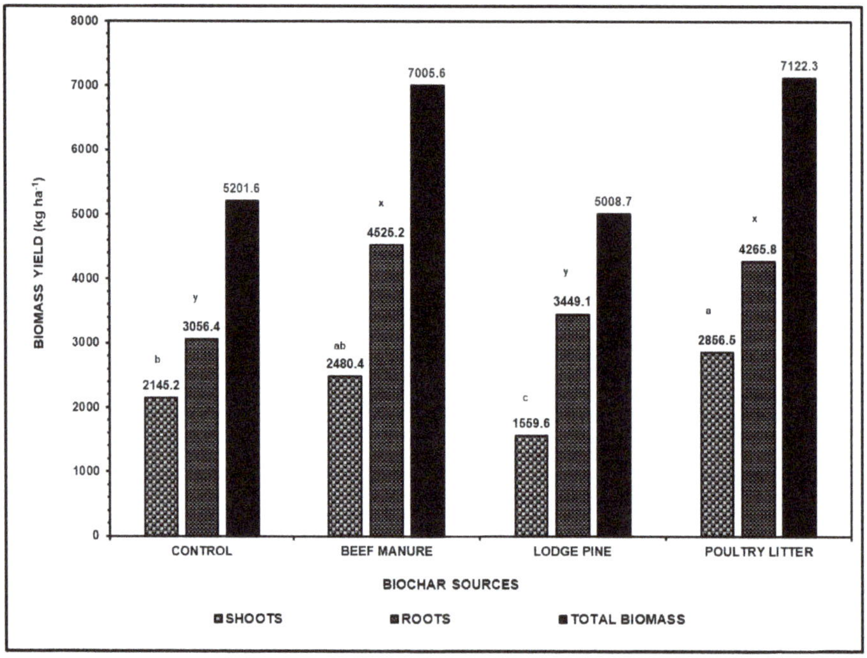

Figure 1. Shoots, roots, and total biomass yield of corn applied with different additions of biochars.

Table 5. Average shoots biomass (SBY) and uptake of Cd and Zn in shoot biomass of corn.

Biochar Additions	Biochar Rate (%)	Compost Rate (%)	SBY (kg/ha)	Cd (kg/ha)	Zn (kg/ha)
Control	0	0	850.3 ± 49.7	18.0 ± 4.9	298.7 ± 86.4
		2.5	2119.1 ± 139.5	30.7 ± 2.9	816.3 ± 74.8
		5.0	3466.1 ± 711.3	33.8 ± 2.9	1079.1 ± 103.4
	Mean		**2145.2 ± 189.6**	**27.5 ± 7.9**	**731.6 ± 352.7**
Beef Cattle Manure	2.5	0	2020.4 ± 428.6	40.4 ± 4.03	979.8 ± 161.5
		2.5	2300.4 ± 506.7	27.8 ± 2.79	817.2 ± 119.3
		5.0	3544.2 ± 225.7	34.2 ± 2.28	961.1±40.3
	Mean		**2621.6 ± 785.0**	**34.1 ± 6.1**	**919.4 ± 128.2**
	5.0	0	2024.8 ± 380.8	35.9 ± 5.4	897.9 ± 163.4
		2.5	2524.4 ± 968.9	24.6 ± 7.4	626.4 ± 215.5
		5.0	2468.4 ± 623.3	17.1 ± 4.4	387.5 ± 91.1
	Mean		**2339.2 ± 651.4**	**25.9 ± 9.7**	**637.3 ± 263.2**
Lodge Pole Pine	2.5	0	781.1 ± 150.5	12.6 ± 7.1	214.4 ± 127.8
		2.5	1427.9 ± 150.5	23.5 ± 6.7	579.0 ± 187.4
		5.0	2220.8 ± 314.9	34.2 ± 2.0	930.6 ± 74.4
	Mean		**1476.6 ± 702.6**	**23.4 ± 10.6**	**574.6 ± 332.3**
	5.0	0	654.3 ± 71.1	13.9 ± 3.0	212.8 ± 39.0
		2.5	1979.1 ± 248.5	32.9 ± 4.2	774.4 ± 97.3
		5.0	2294.5 ± 845.0	31.8 ± 10.6	819.3 ± 91.8
	Mean		**1642.6 ± 873.7**	**26.2 ± 10.9**	**602.2 ± 331.2**
Poultry Litter	2.5	0	2368.2 ± 607.5	54.2 ± 11.2	737.9 ± 174.4
		2.5	3125.7 ± 980.3	49.9 ± 14.6	689.6 ± 181.7
		5.0	3187.5 ± 203.1	40.2 ± 2.8	538.1 ± 82.5
	Mean		**2893.8 ± 706.4**	**48.1 ± 11.2**	**655.2 ± 160.4**
	5.0	0	3242.1 ± 861.6	25.5 ± 7.3	208.4 ± 50.8
		2.5	2766.1 ± 272.5	15.1 ± 1.8	127.9 ± 9.0
		5.0	2449.2 ± 433.1	12.3 ± 0.9	114.7 ± 12.8
	Mean		**2819.1 ± 608.7**	**17.6 ± 7.1**	**150.3 ± 51.3**
Sources of Variation			**Level of Significance**		
Biochar Additions (BA)			***	**	***
Biochar Rate (BR)			ns	***	***
Compost Rate (RMBC)			***	ns	***
BA × BR			ns	***	***
BA × RMBC			**	***	*
BR × RMBC			*	ns	**
BA × BR × RMBC			ns	ns	**

*** Significant at $p \leq 0.0001$; ** Significant at $p \leq 0.001$; * Significant at $p \leq 0.01$; ns – not significant.

Table 6. Average roots biomass (RBY) and uptake of Cd and Zn in root biomass of corn.

Biochar Additions	Biochar Rate (%)	Compost Rate (%)	RBY (kg/ha)	Cd (kg/ha)	Zn (kg/ha)
Control	0	0	2010.7 ± 122.6	45.0 ± 7.5	972.7 ± 94.1
		2.5	2738.1 ± 496.6	71.5 ± 30.5	1025.8 ± 181.8
		5.0	3420.4 ± 456.9	84.2 ± 8.4	1202.8 ± 123.7
	Mean		**3056.4 ± 453.9**	**66.9 ± 23.7**	**1067.1 ± 231.1**
Beef Cattle Manure	2.5	0	3667.9 ± 414.8	99.7 ± 19.2	1615.8 ± 293.9
		2.5	4079.1 ± 592.7	111.8 ± 3.7	1437.8 ± 56.6
		5.0	4292.8 ± 719.2	104.9 ± 27.0	1228.6 ± 224.8
	Mean		**4013.3 ± 579.5**	**105.5 ± 17.5**	**1427.4 ± 251.4**
	5.0	0	4211.7 ± 210.1	104.9 ± 27.0	1570.3 ± 169.6
		2.5	5570.5 ± 840.5	119.5 ± 23.6	1493.9 ± 241.6
		5.0	5328.8 ± 179.7	120.2 ± 15.3	1101.3 ± 339.8
	Mean		**5036.9 ± 964.2**	**114.0 ± 25.5**	**1388.5 ± 313.3**
Lodge Pole Pine	2.5	0	2586.3 ± 180.1	39.4 ± 11.9	695.1 ± 186.9
		2.5	2670.3 ± 338.2	61.9 ± 26.6	887.2 ± 301.6
		5.0	4723.1 ± 989.8	108.4 ± 23.6	1473.5 ± 367.1
	Mean		**3326.6 ± 174.7**	**69.9 ± 35.8**	**1018.6 ± 434.2**
	5.0	0	2125.0 ± 310.2	32.5 ± 6.3	631.8 ± 197.2
		2.5	3547.1 ± 263.2	60.4 ± 99.6	1051.4 ± 69.5
		5.0	5042.9 ± 806.2	99.6 ± 37.2	1394.7 ± 648.8
	Mean		**3571.7 ± 189.2**	**64.2 ± 34.9**	**1025.9 ± 475.1**
Poultry Litter	2.5	0	4195.5 ± 864.4	93.8 ± 13.9	931.3 ± 202.9
		2.5	3704.8 ± 610.5	97.6 ± 40.5	783.5 ± 76.7
		5.0	4141.0 ± 994.6	67.4 ± 24.4	799.3 ± 298.3
	Mean		**4013.8 ± 762.6**	**86.3 ± 28.4**	**838.0 ± 212.8**
	5.0	0	5832.8 ± 604.9	52.3 ± 20.5	588.3 ± 246.7
		2.5	3765.2 ± 668.6	19.6 ± 5.2	236.6 ± 71.4
		5.0	3955.3 ± 488.8	21.1 ± 3.6	259.5 ± 36.6
	Mean		**4517.8 ± 339.7**	**31.1 ± 19.2**	**361.4 ± 214.2**
Sources of Variation				**Level of Significance**	
Biochar Additions (BA)			***	***	***
Biochar Rate (BR)			*	**	*
Compost Rate (RMBC)			*	ns	ns
BA × BR			ns	**	*
BA × RMBC			**	**	*
BR × RMBC			ns	ns	ns
BA × BR × RMBC			ns	ns	ns

*** Significant at $p \leq 0.0001$; ** Significant at $p \leq 0.001$; * Significant at $p \leq 0.01$; ns – not significant.

The effect of BA on the shoot biomass (kg ha^{-1}) is as follows: PL (2856.6) > BCM (2480.4) > Control (2145.2) > LPP (1559.6) while the effect of BA on the root biomass is the following: PL (4265.8) > BCM (4525.2) > LPP (3449.1) > Control (3056.4). The mean shoot biomass (kg ha^{-1}) of corn following application of 2.5% BCM was about 2621.6 ± 785.0 compared with 2339.2 ± 651.4 from corn treated with 5% BCM. The application of 2.5% LPP and 5% LPP resulted in 1476.6 ± 702 and 1642.6 ± 873.7 while the application of 2.5% PL and 5% PL resulted in 2893.8 ± 706.4 and 2819.1 ± 608.7 kg ha^{-1} of the shoots biomass (Table 5). The effect of increasing rates of the beef manure biochar was more significant because of the increasing trend in the root biomass.

The application of 2.5% LPP and 5% LPP resulted in 3326.6 ± 174.7 and 3571.7 ± 189.2 while the application of 2.5% PL and 5% PL resulted in 413.8 ± 762.6 and 4517.8 ± 339.7 kg ha^{-1} of the roots biomass. The mean corn root biomass (kg ha^{-1}) following the application of 2.5% BCM was about 4013.3 ± 579.5 compared with 5036.9 ± 964.2 from corn treated with 5% BCM. These roots biomass following the application of 2.5% and 5% BCM, 2.5% and 5% LPP, and 2.5% and 5% PL were 31.3% and 64.8%, 8.8% and 16.8%, and 31.3% and 47.8% more when compared with the root biomass from the untreated corn plants, respectively (Table 6). Overall, our results show the beneficial effects of biochars in combination with the compost on enhancing the shoot and root biomass of corn grown in this mine soil.

3.4. Uptake and Bioconcentration Factor of Cd and Zn by Shoots and Roots of Corn

Except for LPP, all applications of biochars had significantly enhanced the shoot uptake of Cd and Zn when compared to the Cd and Zn uptake of untreated corn (Table 5). Similarly, all applications of biochar had significantly enhanced the root uptake of Cd and Zn, except for LPP when compared with the Cd and Zn uptake of the control plants (Table 6). Compared to the shoot uptake (kg ha^{-1}) of Cd and Zn by the control plants of 18.0 ± 4.9 and 298.7 ± 86.4, the application of BCM, LPP, and PL resulted in an average increase of the Cd shoot uptake of 112.2%, −26.7%, and 121.7% and Zn shoot uptake of 214.3%, −46.3%, and 58.8%, respectively (Table 5). On the root uptake of Cd and Zn, the application of BCM, LPP, and PL resulted in 127.3% and 63.8%, −20.2% and −31.8%, 62.4% and −21.9% over the untreated plants, respectively (Table 6). These results suggest that the effects of biochar application on the shoot and root uptake of Cd and Zn by corn may vary significantly with biochars produced from different feedstocks.

The interaction effects of BA × BR × RMBC did not affect the shoot and root uptake of Cd and Zn by corn (Tables 5 and 6). However, the shoot uptake of Zn by corn varied significantly with the interaction of BA × BR × RMBC. The greatest shoot uptake of Zn was from corn plants treated with 2.5% BCM while the least amount of the Zn shoot uptake was from plants applied with 5% PL in combination with 5% raw beef manure. The shoot and root uptake of Cd and Zn by corn varied significantly with the interaction effects of BA × BR (Tables 5 and 6). The greatest shoot uptake of Cd (48.1 kg ha^{-1}) was from plant treated with 2.5% PL while the least amount of the Cd shoot uptake was from plants treated with 5% PL. The application of 5% BCM resulted in the greatest root uptake of Cd (114.1 kg ha^{-1}) while the application of 5% PL had the least amount of Cd root uptake of 31.1 kg ha^{-1}. Corn plants treated with 2.5% BCM (919.4 kg ha^{-1}) had the greatest shoot uptake of Zn while the least Zn shoot uptake by corn was from the application of 5% PL with mean value of 150.3 kg ha^{-1}. Similarly, the greatest Zn root uptake of 1427.4 kg ha^{-1} was from corn treated with 2.5% BCM and the least amount of root uptake of Zn was from plants applied with 5% PL with mean uptake of 361.4 kg ha^{-1}. Our results suggest that corn is an efficient plant in phytostabilizing Cd and Zn when applied with 2.5% biochar with or without compost.

The bioconcentration factor or BCF of Cd and Zn, which is related to the shoot and root uptake of Cd and Zn as affected by BA and BR as shown in Table 7. When averaged across BR, the greatest BCF for Cd was in the shoot of 92.28 due to the application of BCM and the least BCF was from the application of PL (72.81). The BCF for Zn in the shoot is in the order: BCM (71.88) > LPP (55.10) > PL (35.30). Similarly, both the Cd and Zn BCF in the roots are in the order: BCM (187.80 and 70.39) > LPP (90.54 and 45.08) > PL (83.40 and 40.76), respectively (Table 7). These results suggest a beneficial effect of biochar application in enhancing the phytostabilization capacity of corn roots and shoots for Cd and Zn.

Table 7. Bioconcentration factor of Cd and Zn in corn as affected by different biochar additions and rates of biochar application.

Biochar Additions	Biochar Rate (%)	Cd	Zn	Cd	Zn
		Shoots		Roots	
Beef Cattle Manure Beef	2.5	99.81	74.99	187.65	72.54
	5.0	94.75	68.76	187.95	68.24
Mean		**92.28**	**71.88**	**187.80**	**70.39**
Lodge Pole Pine Beef	2.5	75.16	54.00	95.26	44.42
	5.0	83.50	56.20	85.82	45.75
Mean		**79.39**	**55.10**	**90.54**	**45.08**
Poultry Litter Beef	2.5	76.07	34.82	94.67	30.77
	5.0	69.55	35.78	72.13	50.75
Mean		***72.81***	***35.30***	***83.40***	***40.76***

4. Discussion

Overall, our results showed that mine spoil remediation can be potentially enhanced by using soil amendments capable of immobilizing metal(loid)s when combined with plant species that are tolerant of high levels of contaminants (Table 1). The incorporation of organic amendments improves the quality of mine soils and makes it possible for vegetation to be established [20,21]. Hossain et al. [24] and Dede et al. [26] have reported that the addition of organic amendments (e.g., biochars, sewage sludge, manures) to soil have promoted the phytoextraction process and improved soil conditions to raise the agronomic values of the soils.

Our results validate the beneficial effects of biochars in combination with the beef cattle manure compost on enhancing the shoot and root biomass and nutritional uptake of corn grown in mine soil with heavy metal contaminations. The greatest total corn biomass was from soils treated with manure-based biochars (PL and BCM) and the least total biomass was from wood-based biochar (LPP) untreated soils. The shoot and root biomass varied significantly with different biochar additions. Results have suggested that biochar applications in mine soils are more likely to influence the biomass, and the effect could be long lasting. Several factors could have had affected the outcome of our study. For instance, differences in the rapidity of decomposition and chemical stability between manure-based and wood-based biochars. In addition, the C:N ratio of the biochars, age of feedstocks, and the degree of disintegration or particle size of the biochars can govern the amount of nutrients released in the soil [33,34]. The C:N ratio of the different biochars that were used in the study are as follows: Poultry litter (8.9) < beef cattle manure (13.8) < lodgepole pine (129.3). Lodgepole pine with wide C:N ratio and low nitrogen content (Table 1) is associated with a slow decay while PL and BCM with narrow C:N ratio and containing higher nitrogen content may undergo rapid mineralization. The profound differences in the C:N ratio of these biochars can explain the striking difference in the decomposition rates, hence faster release of nutrients from these additions to the soils. The rates of mineralization in biochars may have had significant effects on the biomass and nutrient uptake of crop. Our results confirmed the significant effects of different additions of biochars with or without beef cattle manures on biomass productivity and Cd and Zn uptake of corn. As observed in our study, improvements in the corn biomass yield after the biochar addition is often attributed to increased water and nutrient retention, improved biological properties and CEC and improvements in soil pH.

Manure-based biochars, particularly when pyrolyzed at higher temperatures (500 °C and above), have been shown to have strong metal binding capabilities [35]; results which are supported by this study with concentrations of water-soluble Cd and Zn lowest in soils amended with both manure-based biochars (PL and BCM). Concomitantly, additions of PL and BCM resulted in increased total plant biomass yields as compared with the untreated soils and wood-based biochar amendments (PLL). These results are potentially indicative of reduced plant toxicity, though another possibility is that reductions in the available soil of Zn and Cd resulted in reduced stress on soil rhizosphere communities.

Rhizospheric microbial communities provide critical ecosystem services, including nutrient cycling and uptake [36], which result in increased soil fertility. Ippolito et al. [37] previously demonstrated that heavy metal concentrations can have a deleterious effect on microbial community diversity, and additional studies have shown reductions in microbial abundance when faced with increased soil heavy metal concentrations, both of which can negatively impact soil health.

The use of biochar has been investigated for in situ remediation of contaminated lands associated with plants [38,39]. Our results suggest that the incorporation of biochar enhanced phytostabilization of Cd and Zn with concentrations of water-soluble Cd and Zn lowest in soils amended with both manure-based biochars (PL and BCM) while improving the biomass productivity of corn. The biochar application has been shown to be effective in metal immobilization, thereby reducing the bioavailability and phytotoxicity of heavy metals. They also reported that the addition of biochars improve agronomic properties by increasing nutrient availability and microbial activity. The uptake of heavy metals by most plant species decreases in the presence of biochars [40–42]. Further benefits of adding biochars to soil have also been reported; these include the adsorption of dissolved organic carbon [43], increases in soil pH and key soil macro-elements [44], and reductions in trace metals in leachates. Our results support the idea that biochar has proven to be effective at reducing the high concentration of soluble Cd and Zn originating from a contaminated soil and we can now more affirmatively say that sorption is one of the mechanisms by which those metals are retained [45].

The concentrations of water-soluble Cd and Zn in the soil treated with 2.5% and 5% biochars in combination with the increasing beef cattle manure were considerably lower when compared with the control. These results showed effective lowering of Cd and Zn in mine soils after harvesting of corn may well relate to soil pH and phytostabilization of Cd and Zn due to the application of different additions of biochars, especially the manure-based biochar. Sorption of Cd and Zn in biochars can be due to complexation of the heavy metals with different functional groups present in the biochar, such as Ca^{+2} and Mg^{+2} [46], K^{+}, Na^{+} and S [47], or due to physical adsorption [47]. Some other compounds present in the ash, such as carbonates, phosphates or sulphates [48,49] can also help to stabilize heavy metals by precipitation of these compounds with heavy metals [13].

Overall, the pH of mine soils was significantly affected by the increasing rate (2.5% to 5.0%) of different additions of biochars. The soil pH of mine soil treated with 2.5% and 5.0% BCM was increased from 5.2 to 5.61. Similarly, the pH of soils treated with 2% and 5% LPP was increased slightly from 4.7 to 4.8. A much higher increase in the pH of mine soils with 5% PL (6.5) when compared with the control. The application of biochar in our study increased the soil pH and thus enhanced the phytostabilization of metals and our results agreed with the findings of Park et al. [49] and Zhang et al. [50]. The specific mechanism of metal immobilization in the biochar treatments, with increased soil pH, was likely a result in the formation of precipitates such as $Cd(OH)_2$ and $Zn(OH)_2$. For Cd and Zn, the speciation of which in soil solution is more dominated by free metal ion. Shuman [51] reported that at a pH above eight, chemical precipitation took place and therefore retention of Zn in the soil was due to fixation as a solid phase. Singh and Abrol [52] also concluded that above pH 7.9, pH-pZn curves for different soil systems merged and precipitation reactions were controlling Zn retention.

Metal adsorption in the soil, in addition to pH, organic matter has overriding importance on metal solubility and retention in many soils [53]. Few reports in the literature about soil amendments, such as lime and compost being used to reduce the bioavailability of heavy metals [54]. Biochars can also stabilize heavy metals in soils and thus reduce plant uptake [13]. Addition of soil organic matter in the form of BCM has been recognized as a critical component in the retention of heavy metals in our study. For example, soils treated with 5% BA (PL, BCM, or LPP) when combined with 5% BCM had the lowest concentrations of water-soluble Cd and Zn in the soil. A decreasing trend was noted on the concentrations of water-soluble Cd and Zn in soils with increasing rates of the manure-based compost. The addition of MBC may have enhanced the redistribution of Cd and Zn fractions in the soils and enhanced the phytostabilization and bioavailability of these metals [55]. Our results showed that heavy metal concentrations of Cd and Zn in the plants could be profoundly affected by the amount of plant available heavy metals in the soil. Additionally, it is possible that the increase in soil pH caused by the biochar application could have had enhanced the adsorption and complexation of Cd and Zn on biochar, which caused a decrease in water-soluble Cd and Zn in the soil at 5% level of biochars in our study. It has been shown that organic materials can strongly bind heavy metals such as Cu, Pb, Cd, Zn, and Ni. The solubility of the metals depends mainly on the metal loading over soil sorbents, pH, and the concentration of dissolved organic matter in the soil solution [56].

Another important part of this study is on the effect of different additions and application rates of biochars on the bioconcentration factor (BCF) of Cd and Zn in corn shoots and roots. Plant's ability to accumulate metals from soils can be estimated using BCF, which is defined as the ratio of metal concentration in the shoots or roots to that in the soil. The plant's ability to translocate metals from the roots to the shoots is measured using the translocation factor (TF), which is defined as the ratio of the metal concentration in the shoots to the roots. As shown in our data (Table 7), corn has demonstrated a high degree of tolerance factor because we did not see restriction in soil-root and root-shoot transfers. Corn grown in contaminated mine soils can be considered as a hyperaccumulator because it has actively taken up and translocated Cd and Zn into their biomass. Our results showed that BCF of Cd and Zn varied significantly with the different additions and application rates of biochars. Corn applied with 2.5% BCM has the greatest Cd and Zn BCF in the shoots and these results suggest that corn can accumulate large quantities of metal in their shoot tissues when grown in contaminated mine soils. Based on averaged BCF in corn with different additions and rates of biochars, corn can be considered a minor accumulator of Cd and Zn. However, the BCF values of Cd and Zn in corn (Table 7) were much greater than one, are evident that Cd and Zn in mine soils were highly bio-accumulated and phytostablized. Lu et al. [57] from their study on the removal of Cd and Zn by water hyacinth suggested that water hyacinth as a moderate accumulator of Cd and Zn with BCF values of 622 and 789, respectively. Another study on the use of biochar and phytostabilization using *Brassica napus* L. was conducted to target Cd-polluted soils [7]. Additionally, the results of Hartley et al. [58] and Case et al. [59] showed that biochar can be used in combination with Miscanthus for phytostabilization of Cd and Zn in contaminated soils. Novak et al. [60] from their most recent study on using blends of compost and biochars concluded that the designer biochar is an important management component in developing successful mine site phytostabilization program.

5. Summary and Conclusions

In our study, we evaluated the interactive effects of manure- and plant-based biochar applications with or without compost on the shoots and roots biomass production, uptake, and BCF of Zn and Cd of corn grown in mine soil. Results of our study can be summarized as follows:

1. With increasing rates of biochar in combination with increasing rates the application of manure-based compost enhanced soil pH and decreased the concentrations of water-soluble Cd and Zn in mine soils;
2. Effects of the biochar application on the shoot and root uptake of Cd and Zn by corn varied significantly with biochars produced from different feedstocks; and
3. The BCF values of Cd and Zn in corn were considerably greater than one, which are evident that Cd and Zn in mine soils were highly bio-accumulated and phytostablized due to biochar and phytostabilization using corn.

Overall, our results suggest that phytostabilization when combined with the biochar and manure-based compost application have the potential for the remediation of heavy metals polluted soils.

Author Contributions: All authors contributed to this research project. Individual contributions to the following categories are as follows: Research Conceptualization: G.C.S., J.M.N., M.G.J., J.I., T.D.D., and K.S.; Methodology: J.M.N., G.C.S., T.D.D., and D.W.W.; Data Analysis: G.C.S.; Writing—Original draft preparation: G.C.S.; Review and editing: J.M.N., J.I., M.G.J., K.S., T.D.D., and D.W.

Funding: This research was funded through an Interagency Agreement between the United States Department of Agriculture-Agricultural Research Service (60-6657-1-204) and the United States Environmental Protection Agency (DE-12-92342301-2).

Acknowledgments: Gratitude is expressed to the staff of the ARS especially Mr. William Myers, the US EPA locations, and team at the Webb City, MO water treatment facility for their work and diligence with sample collection, preparation and analyses. This work was made possible through an Interagency Agreement between the United States Department of Agriculture-Agricultural Research Service (60-6657-1-204) and the US EPA (DE-12-92342301-2). Approval does not signify that the contents reflect the views of the USDA-ARS or the US EPA, nor does mention of trade names or commercial products constitute endorsement or recommendation for their use. USDA is an equal opportunity provider and employer.

Conflicts of Interest: There is no conflict of interests.

References

1. Lebrun, M.; Macri, C.; Miard, F.; Hattab-Hambli, N.; Motelica-Heino, M.; Morabito, D.; Bourgerie, S. Effect of biochar amendements on As and Pb mobility and phytoavailability in contaminated mine technosols phytoremediated by Salix. *J. Geochem. Explor.* **2017**, *182*, 149–156. [CrossRef]
2. Rodriguez-Vila, A.; Covelo, E.F.; Forjan, R.; Asensio, V. Recovering a copper mine soil using organic amendments and phytomanagement with *Brassica juncea* L. *J. Environ. Manag.* **2015**, *147*, 73–80. [CrossRef] [PubMed]
3. Vega, F.A.; Covelo, E.F.; Andrade, M.L. Limiting factors for reforestation on mine spoil from Galicia (Spain). *Land Degrad. Dev.* **2005**, *16*, 27–36. [CrossRef]
4. Puga, A.P.; Melo, L.C.A.; De Abreau, A.; Coscione, A.R.; Paz-Ferreiro, J. Leaching and fractionation of heavy metals in mining soils amended with biochar. *Soil Tillage Res.* **2016**, *164*, 25–33. [CrossRef]
5. Kloss, S.; Zehetner, F.; Oburger, E.; Buecker, J.; Kitzler, B.; Wenzel, W.W.; Wimmer, B.; Soja, G. Trace element concentration in leachates and mustard plant tissue (*Sinapis alba*, L) after biochar application to temperature soils. *Sci. Total Environ.* **2014**, *481*, 498–508. [CrossRef] [PubMed]
6. Ali, H.; Khan, E.; Muhammad, A.S. Phytoremediation of heavy metals—Concept and applications. *Chemosphere* **2013**, *91*, 869–881. [CrossRef] [PubMed]
7. Houben, D.; Evrard, L.; Sonnet, P. Beneficial effects of biochar application to contaminated soils on the bioavailability of Cd, Pb, and Zn and the biomass production of rapeseed (*Brassica napus* L.). *Biomass Bioenergy* **2013**, *57*, 196–2014. [CrossRef]
8. Conesa, H.M.; Evangelou, M.W.H.; Robinson, B.H.; Schulin, R.A. A critical view of current state of phytotechnology to remediate soils: Still a promising tool? *Sci. World J.* **2012**, *2012*, 173829. [CrossRef] [PubMed]
9. Arthur, E.L.; Rice, P.J.; Anderson, T.A.; Baladi, S.M.; Henderson, K.L.D. Phytoremediation-an overview. *Crit. Rev. Plant Sci.* **2005**, *24*, 109–122. [CrossRef]
10. Novak, J.M.; Ippolito, J.A.; Ducey, T.F.; Watts, D.W.; Spokas, K.A.; Trippe, K.M.; Sigua, G.C.; Johnson, M.G. Remediation of an acidic mine spoil: Miscanthus biochar and lime amendment affects metal availability, plant growth, and soil enzyme activity. *Chemosphere* **2018**, *205*, 709–718. [CrossRef]
11. Bolan, N.S.; Park, J.H.; Robinson, B.; Naidu, R.; Huh, K.Y. Phytostabilization: A green approach to contaminant containment. *Adv. Agron.* **2013**, *112*, 145–202.
12. Dudka, S.; Adriano, D.C. Environmental impacts of metal ore mining and processing: A review. *J. Environ. Qual.* **1997**, *26*, 590–602. [CrossRef]
13. Paz-Ferreiro, J.; Lu, H.; Fu, S.; Mendez, A.; Gasco, G. Use of phytoremediation and biochar to remediate heavy metal polluted soils: A review. *Solid Earth* **2014**, *5*, 65–75. [CrossRef]
14. Mendez, M.O.; Maier, R.M. Phytostabilization of mine tailings in arid and semiarid environments—An emerging remediation technology. *Environ. Health Perspect.* **2008**, *116*, 278–283. [CrossRef] [PubMed]
15. Gentcheva-Kostadinova, S.; Zheleva, E.; Petrova, R.; Haigh, M.J. Soil constraints affecting the forest-biological recultivation of coalmine spoil banks in Bulgaria. *Int. J. Surf. Min. Reclam. Environ.* **1994**, *8*, 47–54. [CrossRef]
16. Novak, J.M.; Cantrell, K.B.; Watts, D.W.; Busscher, W.J.; Johnson, M.G. Designing relevant biochar as soil amendments using ligno-cellulosic-based and manure-based feedstocks. *J. Soils Sediments* **2014**, *14*, 330–343. [CrossRef]

17. Rees, F.; Germain, C.; Sterckeman, T.; Morel, L. Plant growth and meatal uptake by non-hyperaccumulating species (*Lolium perenne*) and a Cd-Zn hyperaccumulator (*Noccaea caeruluscens*) in contaminated soils amended with biochar. *Plant Soil* **2015**, *395*, 57–73. [CrossRef]
18. Beesley, L.; Moreno-Jimenez, E.; Gomez-Eyles, J.L.; Harris, E.; Robinson, B.; Sizmur, T. A review of biochars potential role in the remediation, vegetation and restoration of contaminated soils. *Environ. Pollut.* **2011**, *159*, 3269–3282. [CrossRef]
19. Park, J.H.; Choppala, K.G.; Bolan, N.; Chung, J.W.; Chuasavathi, T. Biochar reduces the bioavailability and phytoxicity of heavy metals. *Plant Soil* **2011**, *348*, 439–451. [CrossRef]
20. Asensio, V.; Vega, F.A.; Andrade, M.L.; Covelo, E.F. Technosols made of wastes to improve physico-chemical characteristics of a copper mine soils. *Pedosphere* **2013**, *23*, 1–9. [CrossRef]
21. Sohi, S.P.; Krull, E.; Lopez-Capel, E.; Bol, R. A review of biochars and its use and function in soil. In *Advances in Agronomy*; Sparks, D.L., Ed.; Academic Press: San Diego, CA, USA, 2010; pp. 47–82.
22. Sigua, G.C.; Novak, J.M.; Watts, D.W.; Johnson, M.G.; Spokas, K. Efficacies of designer biochars in improving biomass and nutrient uptake of winter wheat grown in a hard setting subsoil. *Chemosphere* **2016**, *142*, 176–183. [CrossRef] [PubMed]
23. Laird, D.A.; Fleming, P.; Davis, D.D.; Horton, R.; Wang, B.; Karlen, D.L. Impact of biochar amendments on the quality of a typical Midwestern agricultural soil. *Geoderma* **2010**, *158*, 443–449. [CrossRef]
24. Hossain, M.K.; Strezov, V.; Chan, K.Y.; Nelson, P.F. Agronomic properties of wastewater sludge biochar and bioavailability of metals in production of cherry tomato (*Lycopersicon esculentum*). *Chemosphere* **2010**, *78*, 1167–1171. [CrossRef] [PubMed]
25. Asai, H.; Samson, B.K.; Stephan, H.M.; Songyihangsuthor, K.; Homma, K.; Kiyono, Y.; Inoue, Y.; Shiraiwa, T.; Horie, T. Biochar amendment techniques for upland rice production in Northern Laos 1. Soil physical properties, leaf SPAD and grain yield. *Field Crop Res.* **2009**, *111*, 81–84. [CrossRef]
26. Dede, G.; Ozdemir, S.; Hulusi, D.O. Effect of soil amendments on phytoextraction potential of Brassica juncea growing on sewage sludge. *Int. J. Environ Sci. Technol.* **2012**, *9*, 559–564. [CrossRef]
27. Wei, S.; Zhu, J.; Zhou, Q.X.; Zhan, J. Fertilizer amendment for improving the phytoextraction of cadmium by hyperaccumulator Rorippa globose (Turcz). *J. Soils Sediment* **2011**, *11*, 915–922. [CrossRef]
28. Bradford, G.R.; Page, A.L.; Lund, L.J.; Olmstead, W. Trace element concentrations of sewage treatment plant effluents and sludges: Their interactions with soil and uptake by plants. *J. Environ. Qual.* **1975**, *4*, 123–127. [CrossRef]
29. Novak, J.M.; Busscher, W.; Laird, D.L.; Ahmedna, M.A.; Watts, D.A.; Niandou, M.A.S. Impact of biochar amendment on soil fertility of a southeastern coastal plain soil. *Soil Sci.* **2009**, *174*, 105–112. [CrossRef]
30. United States Environmental Protection Agency (US EPA). Method 3050B: Acid Digestion of Sediments, Sludges, and Soils, Part of Test Methods for Evaluating Solid Waste, Physical/Chemical Methods. 1996; SW-846. Available online: https://www.epa.gov/sites/production/files/2015-12/documents/3050.pdf (accessed on 15 May 2018).
31. Hunag, C.Y.L.; Schulte, E.E. Digestion of plant tissue for analysis by ICP emission spectroscopy. *Commun. Soil Sci. Plant Anal.* **1985**, *16*, 943–958. [CrossRef]
32. SAS Institute. *SAS/STAT User's Guide*; Release 6.03; SAS Institute: Cary, NC, USA, 2000.
33. Sigua, G.C.; Novak, J.M.; Watts, D.W.; Cantrell, K.B.; Shumaker, P.D.; Szogi, A.A.; Johnson, M.G. Carbon mineralization in two Ultisols amended with different sources and particle sizes of pyrolyzed biochar. *Chemosphere* **2014**, *103*, 313–321. [CrossRef]
34. Novak, J.M.; Busscher, W.J. Selection and use of designer biochars to improve characteristics of southeastern USA Coastal Plain degraded soil. In *Advanced Biofuels and Bioproducts*; Lee, J.W., Ed.; Springer: New York, NY, USA, 2012; pp. 69–96.
35. Uchimiya, M.; Cantrell, K.B.; Hunt, P.G.; Novak, J.M.; Chang, S. Retention of Heavy Metals in a Typic Kandiudult Amended with Different Manure-based Biochars. *J. Environ. Qual.* **2012**, *41*, 1138–1149. [CrossRef] [PubMed]
36. Van der Heijden, M.G.A.; Bardgett, R.D.; van Straalen, N.M. The unseen majority: Soil microbes as drivers of plant diversity and productivity in terrestrial ecosystems. *Ecol. Lett.* **2008**, *11*, 296–310. [CrossRef] [PubMed]
37. Ippolito, J.A.; Ducey, T.; Tarkalson, D. Copper impacts on corn, soil extractability, and the soil bacterial community. *Soil Sci.* **2010**, *175*, 586–592. [CrossRef]

38. Uchimiya, M.; Lima, I.M.; Klasson, T.; Wartelle, L.H.; Rodgers, J.E. Immobilization of heavy metal ions by broiler litter-derived biochars in water and soil. *J. Agric. Food Chem.* **2010**, *58*, 5538–5544. [CrossRef]
39. Yu, X.Y.; Ying, G.G.; Kookana, R.S. Reduced plant uptake of pesticides with biochar additions to soil. *Chemosphere* **2009**, *76*, 665–671. [CrossRef] [PubMed]
40. Rees, F.; Sterckeman, T.; Morel, L. Root development of non-accumulating and hyperaccumulating plants in metal-contaminated soil amended with biochar. *Chemosphere* **2016**, *142*, 48–55. [CrossRef] [PubMed]
41. Zheng, R.L.; Cai, C.; Liang, J.H.; Huang, Q.; Chen, Z.; Huang, Y.Z.; Arp, H.P.H.; Sun, G.X. The effects of biochar from rice residue on the formation of iron plaque and the accumulation of Cd, Zn, Pb, As in rice (*Oryza sativa* L.) seedlings. *Chemosphere* **2012**, *89*, 856–862. [CrossRef]
42. Karami, N.; Clemente, R.; Moreno-Jimenez, E.; Lepp, N.W.; Beesley, L. Efficiency of green waste compost and biochar amendments for reducing lead and copper mobility and uptake to ryegrass. *J. Hazard. Mater.* **2011**, *191*, 41–48. [CrossRef]
43. Pietikainen, J.; Kiikkila, O.; Fritze, H. Charcoal as a habitat for microbes and its effect on the microbial community of the underlying humus. *Oikos* **2000**, *89*, 231–242. [CrossRef]
44. Beesley, L.; Moreno-Jimenez, E.; Gomez-Eyles, J.L. Effects of biochar and greenwaste compost amendments on mobility, bioavailability and toxicity of inorganic and organic contaminants in a multi-element polluted soil. *Environ. Pollut.* **2010**, *158*, 2228–2287. [CrossRef]
45. Beesley, L.; Marmiroli, M. The immobilization and retention of soluble arsenic, cadmium and zinc by biochar. *Environ. Pollut.* **2011**, *159*, 474–480. [CrossRef] [PubMed]
46. Lu, H.; Zhang, Y.Y.; Huang, X.; Wang, S.; Qiu, R. Relative distribution of Pb^{2+} sorption mechanisms by sludge-derived biochar. *Water Res.* **2012**, *46*, 854–862. [CrossRef] [PubMed]
47. Uchimiya, M.; Chang, S.C.; Klasson, K.T. Screening biochars for heavy metal retention in soils: Role of oxygen functional groups. *J. Hazard. Mater.* **2011**, *190*, 432–444. [CrossRef] [PubMed]
48. Park, J.H.; Choppala, G.H.; Lee, S.J.; Bolan, N.; Chung, J.W.; Edraki, M. Comparative sorption of Pb and Cd by biochars and its implication for metal immobilization in soil. *Water Air Soil Pollut.* **2013**, *224*, 1711–1718. [CrossRef]
49. Cao, X.D.; Ma, L.N.; Gao, B.; Harris, W. Dairy-manure derived biochar effectively sorbs lead and atrazine. *Environ. Sci. Technol.* **2009**, *43*, 3285–3291. [CrossRef] [PubMed]
50. Zhang, X.; Wang, H.; He, L. Using biochars for remediation of soils contaminated with heavy metals and organic pollutants. *Environ. Sci. Pollut. Res.* **2013**, *20*, 8472–8483. [CrossRef] [PubMed]
51. Shuman, L.L. The effect of soil properties on zinc adsorption by soils. *Soil Sci. Soc. Am. J.* **1975**, *43*, 454–458. [CrossRef]
52. Singh, M.V.; Abrol, I.P. Solubility and adsorption of zinc in sodic soils. *Soil Sci.* **1985**, *140*, 406–411. [CrossRef]
53. McBride, M.B.; Richards, B.K.; Steenhuis, T. Bioavailability and crop uptake of trace elements in soil columns amended with sewage sludge products. *Plant Soil* **2004**, *262*, 71–84. [CrossRef]
54. Komarek, M.; Vanek, A.; Ettler, V. Chemical stabilization of metals and arsenic in contaminated soils using oxides-a review. *Environ. Pollut.* **2013**, *172*, 9–22. [CrossRef]
55. Martinez, C.E.; McBride, M.B. Dissolved and labile concentrations of Cd, Cu, Pb, and Zn in aged ferrihydrite-organic matter systems. *Environ. Sci. Technol.* **1999**, *33*, 745–750. [CrossRef]
56. Weng, L.; Temmighoff, E.; Lofts, S.; Tipping, E.; Riemsdijk, W.H. Complexation with dissolved organic matter and solubility control of heavy metals in a sandy soil. *Environ. Sci. Technol.* **2002**, *36*, 4804–4810. [CrossRef] [PubMed]
57. Lu, X.; Kruatrachue, M.; Pkethitiyook, P.; Homyok, K. Removal of cadmium and zinc by water hyacinth, *Eichhornia crassipes*. *Sci. Asia* **2004**, *30*, 93–103. [CrossRef]
58. Hartley, W.; Dickinson, N.M.; Riby, P.; Lepp, N.W. Arsenic mobility in brownfield soil amended with green waste compost or biochar and planted with *Miscanthus*. *Environ. Pollut.* **2009**, *157*, 2654–2662. [CrossRef] [PubMed]

59. Case, S.D.; McNamara, N.P.; Reay, D.S.; Whitaker, J. Can biochar reduce soil greenhouse gas emission from a Miscanthus bioenergy crop? *GCB Bioenergy* **2014**, *6*, 76–89. [CrossRef]
60. Novak, J.M.; Ippolito, J.A.; Watts, D.W.; Sigua, G.C.; Ducey, T.F.; Johnson, M.G. Biochar compost blends switchgrass growth in mine soils by reducing Cd and Zn bioavailability. *Biochar* **2019**, *1*, 97–114. [CrossRef]

environments

MDPI

Article

Designer Biochars Impact on Corn Grain Yields, Biomass Production, and Fertility Properties of a Highly-Weathered Ultisol

Jeffrey M. Novak *, Gilbert C. Sigua, Thomas F. Ducey, Donald W. Watts and Kenneth C. Stone

United States Department of Agriculture, Agricultural Research Service, Coastal Plains, Soil, Water, and Plant Research Center, 2611 West Lucas Street, Florence, SC 29501, USA; gilbert.sigua@ars.usda.gov (G.C.S.); tom.ducey@ars.usda.gov (T.F.D.); don.watts@ars.usda.gov (D.W.W.); ken.stone@ars.usda.gov (K.C.S.)

* Correspondence: Jeff.Novak@usda.gov; Tel.: +1-843-669-5203

Received: 24 April 2019; Accepted: 1 June 2019; Published: 4 June 2019

Abstract: There are mixed reports for biochars' ability to increase corn grain and biomass yields. The objectives of this experiment were to conduct a three-year corn (*Zea mays* L.) grain and biomass production evaluation to determine soil fertility characteristics after designer biochars were applied to a highly weathered Ultisol. The amendments, which consisted of biochars and compost, were produced from 100% pine chips (PC); 100% poultry litter (PL); PC:PL 2:1 blend; PC mixed 2:1 with raw switchgrass (*Panicum virgatum*; rSG) compost; and 100% rSG compost. All treatments were applied at 30,000 kg/ha to a Goldsboro loam sandy (Fine-loamy, siliceous, sub-active, thermic Aquic Paleudult). Annual topsoil samples were collected in 5-cm depth increments (0 to 15-cm deep) and pH was measured along with Mehlich 1 phosphorus (M1 P) and potassium (M1 K) contents. After three years of corn production, there was no significant improvement in the annual mean corn grain or biomass yields. Biochar, which was applied from PL and PC:PL 2:1 blend, significantly increased M1 P and M1 K concentrations down to 10-cm deep, while the other biochar and compost treatments showed mixed results when the soil pH was modified. Our results demonstrated that designer biochar additions did not accompany higher corn grain and biomass productivity.

Keywords: corn production; designer biochars; soil fertility; Ultisol

---

1. Introduction

Biochar is used as an amendment in agricultural soils to improve their physical characteristics [1–4] and to bolster important fertility properties [5–8]. Biochars' ability to improve soil fertility is explained by the composition of organic compounds, which rebuilds soil organic carbon (SOC) levels [9–11] and ash material, which are comprised of important plant macro- and micro-nutrients [12–14].

For farmers and land managers to be able utilize biochar as a soil amendment, there must be a financial realization that crop or biomass yields are significantly improved. Several reviews [15–17] have reported that, while the overall crop productivity improvement is around 10%, positive crop responses are better demonstrated by adding biochar to acidic, nutrient-poor soils in tropical regions, than in soils in temperate regions. Furthermore, the variability of biochar performance for improving crop yields was further demonstrated by Spokas et al. [18], who reported that biochar caused positive yield increases in 50% of examined studies, but in the remaining 50%, there was no improvement or a decrease in crop yields. More recent examples where biochar was applied to field soils, that did not significantly improve corn yields, were reported [19,20]. In these studies, the test sites were in temperate climatic regions that may explain the lack of significant corn grain yield improvements.

The literature has shown that biochar chemical and physical properties can be quite variable because of feedstock choice differences [12,21], pyrolysis temperature selection [22,23], and post production variance in supply chain management and transportation [24]. Thus, biochar variability can cast some confusion on biochar management decisions. Therefore, an alternate paradigm for biochar usage was introduced [6,25], whereby biochar properties would be matched to correct specific soil fertility deficiencies (i.e., low pH, poor plant nutrient levels, etc.). Novak et al. [6] coined the technology "designer biochar". Designer biochars are produced so they have specific chemical (i.e., pH, nutrient contents, etc.) and physical (i.e., pellets, particle size, etc.) properties through the choice of feedstock, pyrolytic temperatures, and biochar morphology [25,26]. This concept has been adjudicated for biochar production by others [27–29], in order to adjust pH in a calcareous soil [7] and raise winter wheat yields [30].

In many other biochar evaluation studies, there is minimal concern of matching the right biochar to the specific soil problem. In contrast to this approach, our study is unique because designer biochars, used in this study, were selected to target specific soil physico-chemical deficiencies. The Goldsboro soil has deficiencies related to crop production that includes, poor water retention, low SOC, and nutrient contents. We based the designer biochar properties on prior laboratory results, using sandy soils that showed improvement in soil nutrient status [6]; SOC contents [31]; moisture retention [32]; and rebalancing P contents in manure-based biochars [26]. Here, designer biochars were created using commercially purchased biochars, produced from poultry litter (PL) and lodgepole pine chip (PC) feedstocks, and in blends using raw (unpyrolyzed) switchgrass compost (rSG). The biochars were mixed with a compost made from raw switchgrass, since other investigations that had used biochar, mixed with compost, found improved corn grain yields in Australia field plots [33] and with improved wheat yields in China [34].

The objectives of this three-year field experiment were to evaluate the effectiveness of these designer biochars by, 1) improving soil fertility characteristics (i.e., pH, soil Mehlich 1 P and K contents), and 2) increasing corn (*Zea mays*, L.) grain yields and biomass production.

2. Materials and Methods

2.1. Site Characteristics and Soil Properties

The 2-ha field site, used in this study, is located on the property of the United States Department of Agriculture-Agricultural Research Service-Coastal Plain Soil, Water, and Plant Research Center, located in Florence, South Carolina, USA (34°14′38″ N and 79°48′45.3″ W). Over 50 years of farming, the field has been under row crop production, including corn (*Zea Mays* L.), soybeans (*Glycine max*), and an assortment of vegetables (i.e., tomato, strawberry, etc.) crops. For production of these crops, the field was cultivated using either, conservation (deep tilled to 30 to 40 cm) and conventional (disking to incorporate surface residue to about 10 cm deep) tillage practices.

Soil in the field is classified as a Goldsboro loamy sand (Fine-loamy, siliceous, sub-active, thermic Aquic Paleudult). The profile characteristics for the Goldsboro series include, a thin Ap horizon (0 to 20 cm deep), and a shallow E horizon (20 to 30-cm below surface) overlying a series of well-developed Bt argillic horizons, that are expressed down to about 165-cm. The C horizon occurs deeper than 165-cm, is sandy to clayey in texture, and contains distinctive redoximorphic features. The series are moderately-well drained and exhibit masses of oxidized yellow and red colored iron mottles/concretions in the Bt horizons. The series occur in the middle coastal plain region and the parent material consists of marine deposits, interlaced with fluvio-marine sediments [35].

2.2. Designer Biochar Preparation and Characterization

The PC and PL biochars were available commercially. Lodgepole pine (*Pinus contorta*) chips were transformed into biochar using a two-stage process as described [36]. In the first stage of the pyrolysis process, the chips were exposed to temperatures between 500 and 700 °C for <1 min under a very low O_2 atmosphere. In the second stage, the chips were furthered carbonized in an anaerobic environment at temperatures between 300 and 550 °C for approximately 15 min, then removed and allowed to air-cool. The PL biochar was produced using a gasification process employing a fixed-bed pyrolizer programed for conditions (temperature and hold time) that are propriety. The switch-grass feedstock was obtained from plots grown at the Clemson University Pee Dee Research and Education Center, Darlington, South Carolina, USA. The switchgrass was processed using a mechanical grinder to produce 6-mm sized flakes. The bulk material was allowed to compost for 1 week prior to application.

All five amendments were characterized for their pH in a 1:2 (w/w) ratio with deionized water [26]. Additionally, all five amendments were characterized by ultimate analysis, using ASTM method D 3176 ([37]; Hazen Research, Inc., Golden, Colorado, USA] for their ash, fixed C, volatile matter, C, H, O, N and S contents (Table 1). Their molar H/C and O/C ratios were calculated from the elemental analysis (Table 1). The total concentrations of P, K, Cu and Zn concentrations were determined on the ash fraction in 100% PC and 100% PL biochar amendments by first ashing the samples at 600 °C, digesting the ash using method SW866 [37], and then quantifying metal content using ICP-OES by Hazen Research, Inc. The P, K, Cu, and Zn concentrations in the switchgrass compost were determined using acid digestion by the Clemson University Agricultural Service Laboratory (https://www.clemson.edu/public/regulatory/ag-srvc-lab).

Table 1. Designer biochar and compost characteristics (PC = pine chip; PL = poultry litter, rSG = raw switchgrass as compost; nd = not determined).

Parameter (%)	100% PC	100% PL	PC:PL 2:1	PC:rSG 2:1	rSG
C	88.5	33.2	74.4	76.8	51.9
H	1.64	2.23	2.51	3.49	5.61
O	5.91	4.21	5.95	16.27	37.7
N	0.49	3.6	1.41	0.34	0.35
S	0.011	2.6	0.48	0.02	0.01
P	0.025 †	3.36 †	nd	nd	0.07
K	0.301 †	7.23 †	nd	nd	0.29
Cu	0.019 †	0.16 †	nd	nd	0.0007
Zn	0.008 †	0.100 †	nd	nd	0.0016
Ash	3.46	54.1	15.2	3.07	4.38
Fixed C	85.7	16.5	67.6	62.7	17.4
Volatile matter	10.8	29.4	17.2	34.2	78.2
pH	7.8	9.1	9.1	7.3	5.4
O/C	0.05	0.094	0.059	0.15	0.54
H/C	0.22	0.81	0.4	0.54	1.29

† determined on an ash-basis.

2.3. Field Plot Establishment, Soil Sampling, Biochar Application, Corn Management, and Precipitation Recordings

In December 2015, twenty-four plots were established in a randomized complete block design, that allowed for n = 4 plots per amendment treatment and a control (no organic amendments). Each plot area covered 40 m^2 that would allow for future planting of 4 rows of corn per plot on a 0.76-m row spacing. In January 2016, soil bulk density samples were collected in 5-cm depth increments to a depth of 15-cm from hand dug, shallow excavation pits at 1 location in each plot. The soils were oven-dried overnight at 105 °C weighed, and their bulk density values calculated accordingly to [38]. For fertility assessment, soil samples were collected at 6–8 randomly selected locations in each plot at 0–5, 5–10, and 10–15-cm depth increments using a 2.5-cm diameter sampling probe. The samples were composited by depth and dried at 105 °C prior to analysis. Soil samples were then analyzed for pH, Mehlich 1 K (M1 K), and Mehlich 1 P (M1 P) by the Clemson University Agricultural Service Laboratory (https://www.clemson.edu/public/regulatory/ag-srvc-lab). Soil fertility concentrations by depth and plot assignment were matched with their respective soil bulk density measurements, and their final values were reported on a kg/ha basis. This soil sampling scheme and fertility assessment was repeated in January 2017 and 2018.

In February 2016, biochar and biochar/compost mixtures were formulated in the field, and were hand-applied and initially raked into the soil. Each plot received the equivalent of 30,000 kg/ha of amendment (Table 2). After hand-application ceased, the amendments were then mixed to 10-cm soil depth by disking with a cultivator.

Table 2. Formulation and application of biochars and compost per treatment.

Treatment	Biochar (kg/ha)	Compost (kg/ha)
Control	0	0
100% pine chip (PC)	30,000	0
100% poultry litter (PL)	30,000	0
PC:PL 2:1 blend	20,000	10,000
PC: raw switchgrass (rSG) 2:1	20,000	10,000
rSG (compost)	0	30,000

In April 2016, inorganic N-P-K starter fertilizer was applied to all plots at described rates (Table 3). During May 2016, a split application of liquid N was made by side dressing along corn rows. Before corn planting, all plots were initially fertilized with a granular fertilizer, containing monoammonium phosphate (MAP) and a potassium source (K_2O). In all three years, corn was planted in April with the same corn variety, and planting rate was maintained (Table 3). A few weeks after planting, corn was fertilized with liquid N in a split application at the rate of 67 kg/ha (total annual N of 147.5 kg/ha applied). Except in 2018, MAP was not applied because soil fertility measurements showed sufficient soil P concentrations, so the total N applied was reduced to 134 kg/ha.

Table 3. Agronomic management of plots.

Year	N-P-K Applied (kg/ha)	Corn Variety	Stand Count (Plants/ha)
2016	13.5-67-84 (starter) 67-0-0 (split) 67-0-0 (split)	DKC64-69	59,406
2017	13.5-67-84 (starter) 67-0-0 (split) 67-0-0 (split)	DKC64-69	59,406
2018	0-56-0 67-0-0 (split) 67-0-0 (split)	DKC64-69	59,406

In 2016, 2017, and 2018, corn was planted during one operation by chiseling tilling the soil down to 40-cm using a KMC deep tiller (Kelly Manufacturing Co., Tifton, Georgia, USA) and the seeds were planted, using a Case model 1210 corn planter (Case, Inc., Grand Island, Nebraska, USA. Corn variety and stand count were kept consistent across the study period (Table 3). At the end of the season, corn grain was harvested using a mechanical harvester on the two center corn rows of each plot. Corn biomass (mass without grain weight) was quantified by physically harvesting a randomly selected 2 m row section of the two middle corn rows. Plant material was placed into burlap sacks and dried at 60 °C, and then weighted. The biomass measured was then calculated on a dry-mass (0% moisture content) kg/ha basis.

The daily precipitation at the location was recorded using a rain gauge. Daily precipitation amounts were then composited for monthly and yearly totals. When precipitation results were not available, due to power outages/damage to equipment from Hurricanes Mathew (2016), and Florence (2018), the rainfall results were used from a nearby weather station at the Pee Dee SCAN site located in Darlington, South Carolina, USA (https://wcc.sc.egov.usda.gov/nwcc/site?sitenum=2037&state=sc).

The cumulative monthly precipitation for the critical corn growth cycle along with the dry corn yield per plot was used to determine the water use efficiency (WUE). The WUE was calculated by dividing the mean plot yields by the cumulative precipitation totals (April to August; [39]). The values of WUE were reported as kg grain per ha/mm precipitation [40].

2.4. Statistics

A 2-way ANOVA was used on the mean dry corn grain yields and dry corn biomass results, with the fixed variables being year of study and biochar treatments, and with a year x biochar treatment interaction. A 2-way ANOVA was also used for soil pH, M1 K, and M1 P concentrations within biochar treatments, using the year of study and topsoil depths as fixed variables and their interaction was determined. The results were compared by soil depth, since plant nutrients are subject to leaching in sandy soils. All statistical analyses were determined using Sigma Stat v. 11 (SSPS Corp., Chicago, IL, USA) at a $P < 0.05$ level of significance.

3. Results

3.1. Designer Biochar Formulations, Application, Agronomic Management, and Precipitation Totals

All plots received a cumulative amount of amendments at 30,000 kg/ha (Table 2). In all three years of this study, the same corn variety was planted, but the fertilizer management varied slightly (Table 3). Annual precipitation totals were quite variable over the three years, with totals ranging from 1116.7 to 2309.5-mm (Table 4).

Table 4. Monthly and annual precipitation totals for study period.

	Precipitation (mm)		
Month	**2016**	**2017**	**2018**
January	57.7	81.3	72.1
February	153.9	48	35.6
March	45.7	50.5	347
April	73.1	104.1	126.4
May	108.1	46.9	129.5
June	96.1	170.1	88.4
July	213.1	199.6	122.4
August	50.8	94.2	66
September	35.1	101.3	382.8
October	322.3	71.9	469.6
November	29.46	17.5	158.2
December	154.9	131.3	311.5
Annual total	1340.3	1116.7	2309.5

3.2. Corn Grain, Biomass Yields, and Water Use Efficiency

The annual mean corn grain yields (at 0% moisture content), sorted by treatment and production year, are presented in Table 5. In 2016, corn yields from all treated plots were not significantly different than the control. It was noted, however, among the treatments in 2016, there was significant differences between only two treatments, with corn grain yields from 100% PL being > PC:rSG 2:1 and PC:PL 2:1 blend being > PC: rSG 2:1 (Table 3). The differences in corn grain yields were >2,300 kg/ha between these treatments. In 2017 and 2018, however, there was no significant corn grain yield differences between all treatments. Additionally, when corn grain yields were average across years, there was no significant differences between treatment means (Table 5).

Table 5. Comparison of annual mean corn grain yields (n = 4; PC = pine chip; PL = poultry litter, rSG = raw switchgrass compost).

	Corn Grain Yields (dry, kg/ha)			
Treatment	**2016** †	**2017**	**2018**	**Mean**
100% PC	10,356 a, A	9130 ab, A	6837 c A	8774 A
100% PL	10,515 a, A	8092 b, A	7205 b, A	8603 A
PC:PL 2:1	10,529 a AB	8506 b, A	6949 c, A	8661 A
PC:rSG 2:1	8153 a, AC	8273 a, A	6913 a, A	7779 A
rSG	9164 a, A	8232 ab, A	6091 b, A	7828 A
Control (0 biochar)	9110 a, A	8518 ab A	6669 b, A	8099 A
mean	9637 a	8459 b	6777 c	
Source of variation	*P*			
Year	<0.001			
Biochar treatment	0.085			
Yr*Biochar trt	0.356			

† lower case letter indicates significant differences among means between columns, while a capital letter indicates significant differences among means within a column using a 2-way ANOVA at $P < 0.05$ level of significance.

When each treatment mean was compared between years of production, 5 out of 6 treatments experienced a significant corn grain yield decline, while only the PC:rSG 2:1 treatment remained similar (Table 5). Comparing the mean corn yields, when averaged across all 6 treatments, revealed that the means were significantly reduced from 9637 in 2016 to 6777 kg/ha in 2018 (Table 5). Corn grain yield differences between 2016 versus 2018, calculated out to a reduction of 2860 kg/ha or about −30% change.

The effects of biochar and compost on corn biomass (without corn grain weight) were measured by year and treatment (Table 6). In 2016, there were more significant differences between treatments than noted in 2017 and 2018. Additionally, significant biomass yield declines were noted in 5 of the 6 treatments (except 100% PC) and when averaged across all treatments by year. In fact, the differences between mean biomass yields, in 2016 and 2017, were 2344, and between 2017, and 2018 were 1265 kg/ha, respectively. Overall, the mean biomass yields across all treatments were significantly lower from 2016 to 2018 with a decline of 3709 kg/ha (Table 6).

Table 6. Comparison of annual mean corn biomass yields (above ground biomass minus corn grain weights; n = 4; PC = pine chip; PL = poultry litter, rSG = raw switchgrass compost).

	Corn Biomass Yields (Dry, kg/ha)			
Biochar Treatment	**2016** †	**2017**	**2018**	**Mean**
100% PC	8365 a, A	7620 a, A	7107 a, A	7787 A
100% PL	11,573 a, B	8723 b, A	7202 b, A	9166 BC
PC:PL 2:1	11,173 a, BC	9145 b, A	6928 c, A	9082 C
PC:rSG 2:1	9947 a, ABC	7018 b, A	7101 b, A	8022 ABC
rSG	11,354 a, BC	7857 b, A	6668 b, A	8626 ABC
Control (0 biochar)	9430 a, ABC	7684 ab, A	5934 b, A	7683 A
mean	10,352 a	8008 b	6823 c	
Source of variation	*P*			
Year	<0.001			
Biochar treatment	0.003			
Yr*Biochar trt	0.143			

† lower case letter indicates significant differences among means between columns, while a capital letter indicates significant differences among means within a column using a 2-way ANOVA at $P < 0.05$ level of significance.

The WUE results calculated for 2016, 2017, and 2018 are presented in Table 7. The highest WUE calculated occurred in 2016, with values ranging from 15.1 to 19.4. The overall annual WUE mean for 2016 is 17.8 but has a sizable standard deviation of 1.8. In 2016, three of the treatments in 2016 (PC:rSG

2:1, rSG and the control) had lower WUE values compared to the other three treatments. The WUE values for the treatments declined in 2017 and 2018 to > 15, compared to the 2016 values. It was also noted that the variability of the mean also declined in 2017 and 2018.

Table 7. Water use efficiency determined for each treatment (units are kg dry corn grain per ha/mm of cumulative monthly precipitation recorded between April to August; PC = pine chip, PL = poultry litter, rSG = raw switchgrass compost; SD = standard deviation).

Biochar trt.	2016	2017	2018
100% PC	19.1	14.8	12.8
100% PL	19.4	13.2	13.5
PC:PL 2:1	19.5	13.8	13
PC:rSG 2:1	15.1	13.5	13
rSG compost	16.9	13.4	11.4
Control (0 biochar)	16.8	13.9	12.5
Overall annual mean (SD)	17.8 (1.8)	13.8 (0.6)	12.7 (0.7)

3.3. Soil Fertility

In the mean annual pH values of the Goldsboro control soil, there were no significant changes between years and depth (Table 8). In contrast, the plots treated with 100% PL biochar had higher annual mean topsoil pH values. In these plots, the annual mean topsoil pH values significantly increased with time and depth ($P < 0.001$ and 0.014, respectively, Table 8). Mean annual topsoil pH values after treatment with 100% PC biochar, the two blends, and the raw switchgrass compost were mixed. While 4 of 6 treatments exhibited significant impacts of time on soil pH, in contrast, only 1 of 6 treatments showed a significant depth impact (Table 9).

The Goldsboro control soil, without the biochar addition, had no significant time or depth effect (Table 9). Meanwhile, amending the Goldsboro soil with 100% PL biochar significantly raised the annual mean topsoil (0 to 15 cm) M1 P concentration. There is a very significant time and depth impact of 100% PL biochar on soil M1 P concentrations ($P < 0.001$; Table 9). In fact, M1 P is stratified in the Goldsboro soils treated with 100% PL biochar because the two topsoil depths had the highest M1 P concentrations, while the 10–15 cm depth soil sample had the lowest M1 P concentration (Table 9). Blending the PL biochar with PC biochar in a 2:1 (w/w) ratio resulted in a decrease in annual mean topsoil M1 P concentrations. There is a very significant impact of time and depth and interaction effect with this treatment ($P < 0.001$). The addition of 100% PC biochar also resulted in small but significant increases in annual mean topsoil M1 P concentrations. In fact, the 100% PC biochar amendments caused both a significant year and depth impact on M1 P concentrations. Finally, the treatment of the Goldsboro soil with PC:rSG 2:1 (w/w) blend and the rSG compost resulted in no effect on soil M1 P concentrations.

The control soils had annual mean topsoil M1 K concentrations that did not vary as a function of time ($P = 0.076$; Table 10). However, there was a significant depth and year * depth interaction ($P < 0.001$), suggesting stratification with higher mean K concentrations in the surface 0–5 cm soil depth. Adding 100% PL biochar, and the two blends significantly increased the soil M1 K concentrations (Table 10). Changes in topsoil M1 K concentrations in the 100% PL treatment is supported by the very significant year effect ($P < 0.001$; Table 10).

Table 8. Mean pH by soil depth per year of study (SOV = source of variation, *P* value in brackets).

Biochar Treatment	Depth (cm)	2016 †	2017	2018	$Mean_{depth}$	SOF (*P*)
Control	0–5	6.52	6.18	6.10	6.27	Year (0.092)
	5–10	6.48	6.35	6.30	6.38	Depth (0.371)
	10–15	6.25	6.25	6.43	6.43	Year * Depth (0.051)
	$Mean_{year}$	6.42	6.25	6.28		
100% pine chip (PC)	0–5	6.43 a, A	6.45 a, A	6.40 a, A	6.43 A	Year (0.004)
	5–10	6.35 a, A	6.53 ab, A	6.83 b, A	6.57 A	Depth (0.125)
	10–15	6.00 a, A	6.40 ab, A	6.67 b, A	6.36 A	Year * Depth (0.090)
	$Mean_{year}$	6.26 a	6.46 ab	6.63 b		
100% poultry litter (PL)	0–5	6.45 a, A	7.03 b, A	6.98 b, A	6.82 A	Year (<0.001)
	5–10	6.38 a, AB	7.03 b, A	7.05 b, A	6.82 A	Depth (0.014)
	10–15	6.10 a, B	6.83 b, A	6.95 b, A	6.63 B	Year * Depth (0.472)
	$Mean_{year}$	6.31 a	6.96 b	6.99 b		
PC:PL 2:1	0–5	6.50	6.65	6.55	6.57	Year (<0.059)
	5–10	6.43	6.78	6.85	6.83	Depth (0.236)
	10–15	6.20	6.38	6.75	6.44	Year * Depth (0.538)
	$Mean_{year}$	6.38	6.60	6.72		
PC:raw switchgrass (rSG) 2:1	0–5	6.53 a, A	6.50 a, A	6.53 a, A	6.52 A	Year (0.001)
	5–10	6.43 a, A	6.50 ab, A	6.78 b, A	6.57 A	Depth (0.098)
	10–15	6.08 a, B	6.45 b, A	6.68 b A	6.40 A	Year * Depth (0.040)
	$Mean_{year}$	6.34 a	6.48 ab	6.66 b		
rSG	0–5	6.38	6.00	5.85	6.08	Year (0.350)
	5–10	6.35	6.05	6.18	6.19	Depth (0.637)
	10–15	5.95	6.05	6.28	6.09	Year * Depth (0.153)
	$Mean_{year}$	6.23	6.03	6.10		

† lower case letter indicates significant differences among means between columns, while a capital letter indicates significant differences among means within a column using a 2-way ANOVA at $P < 0.05$ level of significance.

Table 9. Mean Mehlich 1 P contents (kg/ha) by soil depth per year of study (SOV = source of variation, *P* value in brackets).

Biochar Treatment	Depth (cm)	2016 †	2017	2018	$Mean_{depth}$	SOF (*P*)
Control	0–5	90	128	111	110	Year (0.588)
	5–10	94	101	107	101	Depth (0.227)
	10–15	73	79	84	79	Year * Depth (0.952)
	$Mean_{year}$	86	103	101		
100% pine chip (PC)	0–5	77 a, A	126 b, A	98 ab, A	100 A	Year (0.006)
	5–10	73 a, A	87 a, B	87 a, A	82 B	Depth (<0.001)
	10–15	47 a, A	65 a, B	64 a, A	59 C	Year * Depth (0.358)
	$Mean_{year}$	66 a	93 b	83 ab		
100% poultry litter (PL)	0–5	78 a, A	820 b, A	647 a, A	515 A	Year (<0.001)
	5–10	71 a, A	696 b, A	603 c, A	457 A	Depth (<0.001)
	10–15	54 a, A	210 a, B	207 ab, B	157 B	Year * Depth (<0.001)
	$Mean_{year}$	70 a	576 b	486 c		
PC:PL 2:1	0–5	74 a, A	371 b, A	237 c, A	228 A	Year (<0.001)
	5–10	72 a, A	293 b, B	228 c, A	198 A	Depth (<0.001)
	10–15	54 a, A	129 a, C	123 a, B	102 B	Year * Depth (<0.001)
	$Mean_{year}$	69 a	265 b	196 c		
PC:raw switchgrass (rSG) 2:1	0–5	88 a, A	114 a, A	96 a, A	99 A	Year (0.161)
	5–10	88 a, A	100 a, A	90 a, A	93 A	Depth (0.097)
	10–15	69 a, A	87 a, A	80 a, A	78 A	Year * Depth (0.970)
	$Mean_{year}$	82 a	100 a	89 a		
rSG	0–5	74 a, A	98 a, A	84 a, A	85 A	Year (0.174)
	5–10	70 a, A	79 a, A	76 a, A	75 A	Depth (0.003)
	10–15	42 a, A	62 a, A	52 a, A	52 B	Year * Depth (0.967)
	$Mean_{year}$	62 a	80 a	71 a		

† lower case letter indicates significant differences among means between columns, while a capital letter indicates significant differences among means within a column using a 2-way ANOVA at $P < 0.05$ level of significance.

Table 10. Mean Mehlich 1 K (kg/ha) contents by soil depth per year of study (SOV = source of variation, P value in brackets).

Biochar treatment	Depth (cm)	2016 †	2017	2018	$Mean_{depth}$	SOF (P)
Control	0–5	122 a, A	207 b, A	208 b, A	178 A	Year (0.076)
	5–10	146 a, A	117 a, B	145 a, B	136 B	Depth (<0.001)
	10–15	146 a, A	86 b, B	118 ab, B	117 B	Year * Depth (<0.001)
	$Mean_{year}$	138 a	136 a	157 a		
100% pine chip (PC)	0–5	142 a, A	221 b, A	195 b, A	186 A	Year (0.989)
	5–10	160 a, A	142 a, B	153 a, B	152 B	Depth (<0.001)
	10–15	163 a, A	105 b, C	121 ab, B	130 C	Year * Depth (<0.001)
	$Mean_{year}$	155 a	156 a	156 a		
100% poultry litter (PL)	0–5	137 a, A	494 b, A	329 c, A	320 A	Year (<0.001)
	5–10	161 a, A	524 b, A	310 c, A	332 A	Depth (0.611)
	10–15	161 a, A	462 b, A	268 a, A	297 A	Year * Depth (0.856)
	$Mean_{year}$	153 a	494 b	303 c		
PC:PL 2:1	0–5	125 a, A	331 b, A	250 ab, A	235 A	Year (<0.001)
	5–10	137 a, A	268 a, A	193 a, A	199 A	Depth (0.662)
	10–15	133 a, A	339 b, A	177 b, A	216 A	Year * Depth (0.770)
	$Mean_{year}$	132 a	312 b	207 b		
PC:raw switchgrass (rSG) 2:1	0–5	118 a, A	223 b, A	203 b, A	181 A	Year (0.027)
	5–10	143 a, AB	130 a, B	143 a, B	139 B	Depth (<0.001)
	10–15	148 a, B	104 b, C	113 b, C	122 C	Year * Depth (<0.001)
	$Mean_{year}$	136 a	152 b	153 b		
rSG	0–5	149 a, A	201 a, A	184 a, A	178 A	Year (0.391)
	5–10	164 a, A	127 a, A	128 a, A	140 A	Depth (0.643)
	10–15	154 a, A	301 a, A	106 a, a	187 A	Year * Depth (0.491)
	$Mean_{year}$	156 a	210 a	139 a		

† lower case letter indicates significant differences among means between columns, while a capital letter indicates significant differences among means within a column using a 2-way ANOVA at $P < 0.05$ level of significance.

Likewise, the PC:PL 2:1 blend also increased M1 K extractable concentrations to be approximately 2-fold greater than the control. After some time, however, there was a significant reduction in M1 K ($P < 0.001$; Table 10). There was no depth effect in this treatment. Additions of PC:rSG 2:1 also increased M1 K concentrations but only in the top 0–5 cm soil depth. This stratification contributed to the very significant depth effect and year * depth interaction ($P < 0.001$; Table 10). Finally, rSG additions did not significantly impact soil M1 K concentrations, which is related to its low K concentration in the compost (0.29%; Table 1) and low amount delivered to soil (87 kg/ha; 0.0029 * 30,000 kg/ha; Tables 1 and 2).

4. Discussion

Pine chip biochar was used in this study to bolster the Goldsboro Ap horizon SOC content (0.91% SOC in 0–15 cm deep, data not presented). The 100% PL biochar was selected to bolster the Goldsboro's soils macro (i.e., P and K), and micro (i.e, Cu and Zn) nutrient concentrations. Because PC biochar inherently contains lower quantities of plant nutrients relative to manure-based biochar [14,21], it was blended in a 2:1 (w/w) ratio with nutrient enriched PL biochar (Table 1). This blending of biochars expands the soil fertility benefits by increasing both SOC and plant nutrient concentrations. Additionally, a switchgrass compost was included in the treatments because of anticipated improvement in biochar nutrient transformation processes [33] and soil moisture retention [32,41].

In this study, the biochar applications rates are equivalent to a rate of 30,000 kg/ha (Table 2). This application rate is within the range (10,000 to 50,000 kg/ha) used in other fields [41–43], or in our prior laboratory experiments, involving biochar on sandy coastal plain soils [2,6].

For corn grain yields, the addition of designer biochar to the Goldsboro soil had little influence (Table 5). Some variations between treatments in 2016 did occur, although there was some minor impacts of amendments between the individual treatments. It was noted that, in 2016, corn grain yields from the 100% PL treatment were > PC:rSG 2:1 and the PC:PL 2:1 blend was > PC: rSG 2:1 (Table 5). In term of differences, the corn grain yields were > 2300 kg/ha between these two treatments.

When each treatment mean was compared between years of production, 5 out of 6 treatments experienced a significant corn grain yield decline, while only the PC:rSG 2:1 treatment remained similar. The decline in annual yields was evident by comparing the mean corn yield when averaged across all 6 treatments, which showed that mean corn yields were significantly reduced from 9637 in 2016 to 6777 kg/ha in 2018 (Table 5). Corn grain yield differences between 2016 versus 2018 calculated a reduction of 2860 kg/ha or about −30% change.

The literature has reported mixed results, concerning the biochars' impact on corn grain yields. In a three-year mesocosm experiment, Borchard et al. [19] reported that a wood-based biochar, applied to a sandy Fluvisol and a silty Luvisol, failed to improve corn yields. In a larger field scale study conducted at several United States Department of Agriculture-Agricultural Research Service locations across the USA, Laird et al. [20] reported that a hardwood biochar applied to soils had no significant impact on corn grain yield increases at 5 of the 6 locations. Additionally, Güereña et al. [44] reported that corn yields did not change when grown in two New York soils after biochar additions, even when applied at 30,000 kg/ha. It was speculated in this study that the maize-based biochar did not work in these soils, because there were no fertility constraints, and that the site was in a temperate climate with adequate precipitation totals. In a more recent biochar field study, Lamb et al. [44] also reported no positive impact of a hardwood-derived biochar on corn yields grown in a sandy Ultisol in Georgia.

In contrast, there are numerous biochar studies, conducted under tropical conditions, that have reported maize grain yield increases from field trials, using different feedstocks [45,46]. In one study, Cornelissen et al. [47] reported a positive corn yield increase in a sandy, African Ultisol treated with biochar produced from corn cob/softwood. Additionally, Agegnehu et al. [33] reported a significant corn yield improvement in tropical Ferralsol treated with biochar produced from waste willow wood. These studies reported that biochar has a positive interaction with tropical soils to improve corn grain yields, which was further corroborated in a global-scale meta-analysis, that biochar boosts crop yields in tropical but not temperate zones [17].

The contrasting effects of biochar improving crop yields in tropical soils but having mixed effects at raising yields in temperate soils is a concern. Biochars in tropical soils may be more effective at improving soil fertility conditions by raising low soil pH levels, sequestering phytotoxic aluminum concentrations, adding critical plant nutrients, or by enhancing nutrient turnover properties through stimulating soil microbial populations. Furthermore, biochars' positive crop yield effects, in tropical soils, may be enhanced by mixing with compost [33,34]. On the other hand, biochars' inconsistent performance at increasing crop yields in temperate regions may be related to the wrong biochar applied to an incorrect soil, the soil did not need biochar addition, or that the background soil fertility properties were of sufficient quality to mask biochar responses on soil properties. The ability to explain why crop yields vary with biochar applications under different climate condition or soil properties is problematic. It may be that strategies to improve biochars' inconsistent performance in soils, under temperate climates, will require additional field investigations that specifically identify which soil properties were modified, and how strongly do these changes induce a positive crop yield increase. Otherwise, if biochars expenses are not recouped through associate higher crop grain or biomass yields, then their future use in agricultural as a soil amendment may be limited.

In our study, the effects of biochar and compost on the mean corn biomass (without corn grain weight) showed some significant differences in the first year of study, but were not apparent by the second and third year (Table 6). Similar to the corn grain yields, biomass yields also experienced a significant decline in five of the six treatments (except 100% PC), when averaged across all treatments by year. This represents a 36% decline in mean corn biomass yields, when averaged across all treatments in just three years. This result is similar to corn biomass reductions (i.e., 36%) as reported [48].

Large variations in annual rainfed corn grain yields in the Southeastern USA Coastal Plain region are not uncommon. Heckman and Kamprath [49] observed large annual variations of between 20 to

50% in corn grain yields in a three-year corn production experiment when grown in a NC Dothan loamy sand. Davis et al. [50] also reported that over 11 years, there were large annual mean variations of between 25 to 50% in a corn production experiment grown in a Tifton loamy sand in GA. They attributed the large variation to differences in monthly rainfall. Similar to this study, a decline in corn grain yields may be due to irregular precipitation timing during critical periods of corn pollination and seed filling stages (April to August) of production (Table 4). Although there were near, or above annual precipitation totals (i.e., 1200-mm; [51]), the irregular monthly precipitation totals in May and August 2017, and June and August 2018, probably impacted corn pollination and eventual seed filling. The decline in WUE, calculated in 2017 and 2018, suggests that the corn crop in all treatments was under moisture stress relative to 2016. While the amount of annual precipitation is important, a more vital component is the timing of that precipitation event during critical corn growth cycles to minimize water limited corn yields [52].

Three important soil fertility characteristics were evaluated in this study including, soil pH, M1 P, and M1 K concentrations. Soils were also collected annually in 5-cm topsoil increments because the lack of mechanical mixing during conservation tillage operations was speculated to cause vertical stratification in nutrient concentrations. Additionally, K is reported to readily leach through sandy coastal plain soils after treatment with PL biochar [53]. Sampling, using this procedure, would allow for the assessment of nutrient vertical stratification and for salt leaching, which may influence soil pH or reductions in nutrient concentrations biding in the topsoil.

Th pH range for soils in the control and with biochar treatment (except soil treated with 100% PL biochar) are well within the soil pH realm considered optimum (e.g., pH 5.5 to 6.5) for nutrient availability in Coastal Plain soils [54]. In soil treated with 100% PL biochar, the increase in pH is not unexpected because 100% PL biochars typically have calcareous pH values, due to high concentrations of residual salts in their ash [6,11] and higher ash contents (Table 1). This condition is also related to the Goldsboro soil having a limited ability to buffer salts contained with the 100% PL biochar [6]. With the use of 100% PL biochar on sandy soils, it is important that resultant soil pH values do not exceed seven, since Fe, Mg, Zn, and other micronutrients become less available for plant uptake [54,55].

Three of the six treatments had a significant impact of time on soil pH (Table 8). This condition is probably related to salts leaching out of the biochar as a function of time, and re-establishing the equilibrium with cations associated on clays and in the soil organic carbon pool. After displacement, the salts would promote alkaline conditions because of the higher dissolved Ca and Mg concentrations. Ranking the 2018 mean annual soil pH values grouped by topsoil depth were 100% PL > PC:PL 2:1 > PC:rSG 2:1 ≥ 100% PC > control > rSG treatment. This corroborates that the calcareous 100% PL and PC:PL 2:1 biochars were more effective at raising pH values in the Goldsboro soil than the other treatments.

Biochars, used as soil amendments, can contribute plant nutrients, such as P and K to bolster the overall soil fertility status [14]. As shown in Table 9, biochars had different capabilities of supplying P to soil. Expressing the relative effectiveness of these biochars and compost to supply M1 P to the Goldboro soil are: 100% PL > PC:PL 2:1 blend > 100% PC > PC:rSG = rSG compost.

According to the recommended levels for agronomic crop growth in Coastal Plain soils, M1 P concentrations presented in Table 9 show that they rank in the high (67 to 112 kg/ha) to very high (+112 kg/ha; [54]) range. Obviously, adding 30,000 kg/ha of 100% PL biochar to the Goldsboro soil grossly increased the M1 P concentrations to be much greater than the highest M1 P level recommended for Coastal Plain sandy soils. The depth stratification of M1 P to about 10 cm is reflective of the biochar being disked incorporated after application.

Potassium is an important plant nutrient because it is involved in many enzymatic functions, regulates electrochemical balances between plant organelles, and contributes to osmotic potential reactions of cells and tissues [56]. Because K is involved in many plant physiological functions, for example, corn can have a high K nutrition requirement ranging from 3.2 to 28 kg/ha/d [57]. The exact K nutrition requirement varies with geographic locations due to differences in planting rate, soil water availability, and production stage of growth [57]. Typically, large fertilizer K_2O rates are applied annually for corn production. For example, K_2O application rates ranged from 167 to 224 kg/ha in a field corn experiment in a SC sandy Coastal Plain soil [58]. However, the actual amount of K_2O applied each year depends on antecedent M1 K soil tests values. For example, soil test M1 potassium concentrations ranges for corn production in SC are low (<80 kg/ha); medium (80 to 175 kg/ha); sufficient (176 to 204 kg/ha); high (205 to 263 kg/ha) and excessive (>263 kg/ha; [58]).

Here, the M1 K levels were measured in the Goldsboro control soil rank in the medium soil test category, thereby suggesting a need to maintain inorganic K_2O fertilizer additions. For the M1 K concentrations, it is interesting that there was no depth effect in the Goldsboro soil treated with 100% PL, PC:PL 2:1 biochar, or rSG compost (Table 10). This may be explained by a better degree of physical mixing in these plots. In contrast, soil in plots treated with 100% PC biochar and PC:rSG 2:1 had significant depth effects with greater concentrations measured at the 0 to 5-cm depth. The may be linked to a relatively poorer degree of physical mixing or to the lack of K released from the cellulosic material. Overall, the application of 30,000 kg/ha of 100% PL biochar increased M1 K concentrations, so that it was in the excessive soil test range.

5. Conclusions

Customizing biochar properties to match specific soil deficiencies was suggested as a more effective paradigm for biochar usage. The use of designer biochars was reported to more effectively increase corn grain yields or biomass production, compared to a non-specific biochar. Here, we report that designer biochars were able to improve important fertility properties (e.g., pH, M1 P, and K) in the sandy Goldsboro soil. In spite of the noted soil fertility improvements, however, corn grain and biomass yields were not significantly raised. In comparison, when averaged by year, annual mean grain yield and biomass production both declined by about 30%. The declines were probably due to weather fluctuations during critical corn growth stages (i.e., fertilization, seed filling). The lack of significant improvement in corn yields in this study, corroborates the results from other biochar field research projects, conducted in temperate regions. In conclusion, despite the Goldsboro soil being extensively weathered, it still possessed sufficient soil fertility traits that, with good agronomic practices and timely rainfall, can produce satisfactory corn yields.

Author Contributions: All authors contributed to this research project. Individual contributions to the following categories are as follows: Conceptualization, J.M.N., G.C.S., and T.F.D.; methodology, J.M.N., D.W.W., G.C.S., T.D.D., and K.C.S.; formal aanalysis, J.M.N. and K.C.S.; writing—original draft preparation, J.M.N.; writing—review and editing, J.M.N., K.C.S., and D.W.W.; visualization, J.M.N.

Funding: This research was funded by Agricultural Research Service, grant number 6082-12630-001-00D.

Acknowledgments: Gratitude is expressed to the technical staff for their work and diligence with sample collection, field preparation, and analyses. This work was made possible through the United States Department of Agriculture-Agricultural Research Service (USDA-ARS) National Program 212 (Soil and Air) Project number 6082-12630-001-00D. It has been subject to peer review by USDA-ARS scientists and approved for journal submission. Approval does not signify that the contents of this paper reflect the views of the USDA-ARS nor does mention of trade names or commercial products constitute endorsement or recommendation for their use. USDA is an equal opportunity provider and employer.

Conflicts of Interest: There are no conflict of interests.

References

1. Busscher, W.J.; Novak, J.M.; Evans, D.E.; Watts, D.W.; Niandou, M.A.S.; Ahmedna, M. Influence of Pecan Biochar on Physical Properties of a Norfolk Loamy Sand. *Soil Sci.* **2010**, *175*, 10–14. [CrossRef]
2. Novak, J.M.; Busscher, W.J.; Watts, D.W.; Amonette, J.E.; Ippolito, J.A.; Lima, I.M.; Gaskin, J.; Das, K.C.; Steiner, C.; Ahmedna, M.; et al. Biochars Impact on Soil-Moisture Storage in an Ultisol and Two Aridisols. *Soil Sci.* **2012**, *177*, 310–320. [CrossRef]
3. Mukherjee, A.; Lal, R. Biochar Impacts on Soil Physical Properties and Greenhouse Gas Emissions. *Agronomy* **2013**, *3*, 313–339. [CrossRef]

4. Blanco-Canqui, H. Biochar and Soil Physical Properties. *Soil Sci. Soc. J.* **2017**, *84*, 687. [CrossRef]
5. Liang, B.; Lehmann, J.; Solomon, D.; Kinyangi, J.; Grossman, J.M.; O'Neill, B.; Skjemstad, J.O.; Thies, J.; Luizão, F.J.; Petersen, J.; et al. Black Carbon Increases Cation Exchange Capacity in Soils. *Soil Sci. Soc. J.* **2006**, *70*, 1719–1730. [CrossRef]
6. Novak, J.M.; Lima, I.; Xing, B.; Gaskin, J.W.; Steiner, C.; Das, K.C.; Ahmedna, M.; Rehrah, D.; Watts, D.W.; Busscher, W.J.; et al. Characterization of designer biochar produced at different temperatures and their effects on a loamy sand. *Ann. Env. Sci.* **2009**, *3*, 195–206.
7. Ippolito, J.; Ducey, T.; Cantrell, K.; Novak, J.; Lentz, R. Designer, acidic biochar influences calcareous soil characteristics. *Chemosphere* **2016**, *142*, 184–191. [CrossRef] [PubMed]
8. Domingues, R.R.; Trugilho, P.F.; Silva, C.A.; De Melo, I.C.N.A.; Melo, L.C.A.; Magriotis, Z.M.; Sánchez-Monedero, M.A. Properties of biochar derived from wood and high-nutrient biomasses with the aim of agronomic and environmental benefits. *PLOS ONE* **2017**, *12*, e0176884. [CrossRef] [PubMed]
9. Kleber, M.; Hockaday, W.; Nico, P.S. Characteristics of biochar: macro-molecular properties. In *Biochar for Environmental Management*, 2nd ed.; Lehmann, J., Joseph, S., Eds.; Earthscan: Routledge, NY, USA, 2015; pp. 111–138.
10. Brassard, P.; Godbout, S.; Raghavan, V. Soil biochar amendment as a climate change mitigation tool: Key parameters and mechanisms involved. *J. Environ. Manag.* **2016**, *181*, 484–497. [CrossRef] [PubMed]
11. Novak, J.M.; Johnson, M.G. Elemental and spectroscopic characterization of low-temperature (350 °C) lignocellulosic- and manure-based designer biochars and their use as soil amendments. In *Biochar and Biomass and Waste*; Ok, Y.S., Tsang, D.C., Bolan, N., Novak, J.M., Eds.; Elsevier Publisher: Cambridge, MA, USA, 2018; pp. 21–36.
12. Cantrell, K.B.; Hunt, P.G.; Uchimiya, M.; Novak, J.M.; Ro, K.S. Impact of pyrolysis temperature and manure source on physiocochemical characteristics of biochar. *Bioresour. Technol.* **2012**, *107*, 419–428. [CrossRef] [PubMed]
13. Biederman, L.A.; Harpole, W.A. Biochar and its effects on plant productivity and nutrient cycling: a meta-analysis. *GCB Bioenergy* **2013**, *5*, 202–214. [CrossRef]
14. Ippolito, J.A.; Spokas, K.A.; Novak, J.M.; Lentz, R.D.; Cantrell, K.B. Biochar elemental composition and factors influencing nutrient retention. In *Biochar for Environmental Management*, 2nd ed.; Lehmann, J., Joseph, S., Eds.; Earthscan: Routledge, NY, USA, 2015; pp. 139–163.
15. Jeffery, S.; Verheijen, F.; Van Der Velde, M.; Bastos, A.C. A quantitative review of the effects of biochar application to soils on crop productivity using meta-analysis. *Agric. Ecosyst.* **2011**, *144*, 175–187. [CrossRef]
16. Crane-Droesch, A.; Torn, M.S.; Abiven, S.; Jeffery, S. Heterogeneous global crop yield response to biochar: a meta-regression analysis. *Environ. Lett.* **2013**, *8*, 044049. [CrossRef]
17. Jeffery, S.; Abalos, D.; Prodana, M.; Bastos, A.; Van Groenigen, J.W.; Hungate, B.; Verheijen, F. Biochar boosts tropical but not temperate crop yields. *Environ. Lett.* **2017**, *12*, 53001. [CrossRef]
18. Spokas, K.; Cantrell, K.B.; Novak, J.M.; Archer, D.W.; Ippolito, J.A.; Collins, H.P.; Boateng, A.A.; Lima, I.M.; Lamb, M.C.; McAloon, A.J.; et al. Biochar: A Synthesis of Its Agronomic Impact beyond Carbon Sequestration. *J. Qual.* **2012**, *41*, 973. [CrossRef] [PubMed]
19. Borchard, N.; Siemens, J.; Ladd, B.; Möller, A.; Amelung, W. Application of biochars to sandy and silty soil failed to increase maize yield under common agricultural practice. *Soil N.a.* **2014**, *144*, 184–194. [CrossRef]
20. Laird, D.; Novak, J.; Collins, H.; Ippolito, J.; Karlen, D.; Lentz, R.; Sistani, K.; Spokas, K.; Van Pelt, R. Multi-year and multi-location soil quality and crop biomass yield responses to hardwood fast pyrolysis biochar. *Geoderma* **2017**, *289*, 46–53. [CrossRef]
21. Novak, J.M.; Cantrell, K.B.; Watts, D.W. Compositional and thermal evaluation of lignocellulosic and poultry litter chars via high and low temperature pyrolysis. *Bioenerg. Res.* **2013**, *6*, 114–130. [CrossRef]
22. Zhang, J.; Liu, J.; Liu, R. Effects of pyrolysis temperature and heating time on biochar obtained from the pyrolysis of straw and lignosulfonate. *Bioresour. Technol.* **2015**, *176*, 288–291. [CrossRef]
23. Suliman, W.; Harsh, J.B.; Abu-Lail, N.I.; Fortuna, A.-M.; Dallmeyer, I.; Garcia-Perez, M. Influence of feedstock source and pyrolysis temperature on biochar bulk and surface properties. *Biomass and Bioenergy* **2016**, *84*, 37–48. [CrossRef]
24. Anderson, N.M.; Bergman, R.D.; Page-Dumroese, D.S. A supply chain approach to biochar systems. In *Biochar: A Regional Supply Chain Approach in View of Climate Change Mitigation*; Bruckman, V., Varol, E.A., Uzan, B., Liu, J., Eds.; Cambridge University Press: Cambridge, UK, 2017; pp. 24–45.

25. Novak, J.M.; Busscher, W.J. Selection and use of designer biochars to improve characteristics of southeastern usa coastal plain degraded soils. In *Advanced Biofuels and Bioproducts*; Springer Nature: New York, NY, USA, 2012; pp. 69–96.
26. Novak, J.M.; Cantrell, K.B.; Watts, D.W.; Busscher, W.J.; Johnson, M.G. Designing relevant biochar as soil amendments using lingo-cellulosic-based and manure-based feedstocks. *J. Soil Sediment.* **2014**, *14*, 330–343. [CrossRef]
27. Atkinson, C.J.; Fitzgerald, J.D.; Hipps, N.A. Potential mechanisms for achieving agricultural benefits from biochar application to temperate soils: A review. *Plant Soil* **2010**, *337*, 1–18. [CrossRef]
28. Joseph, S.; Van Zwieten, L.V.; Cjia, C.; Kimber, S.; Munroe, P.; Lin, Y.; Marjo, C.; Hook, J.; Thomas, T.; Nielsen, S.; et al. Designing specific biochars to address soil constraints: A developing industry. In *Biochar and Soil Biota*; Ladygina, N., Rineau, F., Eds.; CRC Press: Boca Raton, FL, USA, 2010; pp. 165–201.
29. Mandal, S.; Sarkar, B.; Bolan, N.; Novak, J.; Ok, Y.S.; Van Zwieten, L.; Singh, B.P.; Kirkham, M.B.; Choppala, G.; Spokas, K.; et al. Designing advanced biochar products for maximizing greenhouse gas mitigation potential. *Crit. Rev. Environ. Sci. Technol.* **2016**, *46*, 1–35. [CrossRef]
30. Sigua, G.; Novak, J.; Watts, D.; Johnson, M.; Spokas, K. Efficacies of designer biochars in improving biomass and nutrient uptake of winter wheat grown in a hard setting subsoil layer. *Chemosphere* **2016**, *142*, 176–183. [CrossRef] [PubMed]
31. Novak, J.M.; Busscher, W.J.; Laird, D.L.; Ahmedna, M.; Watts, D.W.; Niandou, M.A.S. Impact of Biochar Amendment on Fertility of a Southeastern Coastal Plain Soil. *Soil Sci.* **2009**, *174*, 105–112. [CrossRef]
32. Novak, J.M.; Watts, D.W. Augmenting soil water storage using uncharred switchgrass and pyrolyzed biochars. *Soil Use Manag.* **2013**, *29*, 98–104. [CrossRef]
33. Agegnehu, G.; Bass, A.M.; Nelson, P.N.; Bird, M.I. Benefits of biochar, compost and biochar–compost for soil quality, maize yield and greenhouse gas emissions in a tropical agricultural soil. *Sci. Total. Environ.* **2016**, *543*, 295–306. [CrossRef] [PubMed]
34. Lashari, M.S.; Liu, Y.; Li, L.; Pan, W.; Fu, J.; Pan, G.; Zheng, J.; Zheng, J.; Zhang, X.; Yu, X. Effects of amendments of biochar-manure compost in conjuction with pyroligneous solution on soil quality and wheat yield of a salt-stressed cropland from Central China Great Plain. *Field Crops Res.* **2013**, *144*, 113–118. [CrossRef]
35. Daniels, R.B.; Buol, S.W.; Kleiss, H.J.; Ditzler, C.A. *Soil Systems in North Carolina, Technical Bulletin 314*; North Carolina State University: Raleigh, NC, USA, 1999.
36. A Ippolito, J.; Berry, C.M.; Strawn, D.G.; Novak, J.M.; Levine, J.; Harley, A. Biochars Reduce Mine Land Soil Bioavailable Metals. *J. Qual.* **2017**, *46*, 411–419. [CrossRef]
37. ASTM (American Society of Testing and Materials). *Petroleum Products, Lubricants, and Fossil Fuels: Gaseous Fuels, Coal, and Coke*; ASTM International: Conshohocken, PA, USA, 2016.
38. Grossmann, R.B.; Reinsch, T.G. Bulk density and linear extensibility: Core method. In *Methods of Soil Analysis:Part 4 Physical methods, SSSA Book Series 5.4*; Dane, J.H., Topp, G.C., Eds.; Soil Science Society of America: Madison, WI, USA, 2002; pp. 208–228.
39. Stone, K.C.; Bauer, P.J.; Sigua, G.C. Irrigation Management Using an Expert System, Soil Water Potentials, and Vegetative Indices for Spatial Applications. *Trans. ASABE* **2016**, *59*, 941–948.
40. Howell, T.A. Enhancing Water Use Efficiency in Irrigated Agriculture. *Agron. J.* **2001**, *93*, 281. [CrossRef]
41. Zainul, A.; Koyro, H.-W.; Huchzermeyer, B.; Gul, B.; Khan, M.A. Impact of a biochar or a compost-biochar mixture on water relation, nutrient uptake and phytosynthesis of Phragmites karka. *Pedosphere* **2017**. [CrossRef]
42. Hansen, V.; Hauggaard-Nielsen, H.; Petersen, C.T.; Mikkelsen, T.N.; Müller-Stöver, D. Effects of gasification biochar on plant-available water capacity and plant growth in two contrasting soil types. *Soil N.a.* **2016**, *161*, 1–9. [CrossRef]
43. Güereña, D.; Lehmann, J.; Hanley, K.; Enders, A.; Hyland, C.; Rhia, S. Nitrogen dynamics following field application of biochar in a temperate North American maize-based production system. *Plant Soil* **2013**, *365*, 239–254. [CrossRef]
44. Lamb, M.C.; Sorensen, R.B.; Butts, C.L. Crop response to biochar under differing irrigation levels in the southeastern USA. *J. Improv.* **2018**, *32*, 305–317. [CrossRef]
45. Major, J.; Rondón, M.; Molina, D.; Riha, S.J.; Lehmann, J. Maize yield and nutrition during 4 years after biochar application to a Colombian savanna oxisol. *N.a. Soil* **2010**, *333*, 117–128. [CrossRef]

46. Van Zwieton, L.; Kimber, S.; Morris, S.; Chan, K.Y.; Downie, A.; Rust, J.; Joseph, S.; Cowie, A. Effects of biochar from slow pyrolysis of papermill waste on agronomic performance and soil fertility. *Plant Soil* **2010**, *327*, 235–246. [CrossRef]
47. Cornelissen, G.; Martinsen, V.; Shitumbanuma, V.; Alling, V.; Breedveld, G.D.; Rutherford, D.W.; Sparrevik, M.; Hale, S.E.; Obia, A.; Mulder, J. Biochar effect on maize yield and soil characteristics in five conservation farming sites in Zambia. *Agronomy* **2013**, *3*, 256–274. [CrossRef]
48. Lentz, R.D.; Ippolito, J.A. Biochar and Manure Affect Calcareous Soil and Corn Silage Nutrient Concentrations and Uptake. *J. Qual.* **2012**, *41*, 1033. [CrossRef] [PubMed]
49. Heckman, J.R.; Kamprath, E.J. Potassium Accumulation and Corn Yield Related to Potassium Fertilizer Rate and Placement. *Soil Sci. Soc. J.* **1992**, *56*, 141. [CrossRef]
50. Davis, J.G.; Walker, M.E.; Parker, M.B.; Mullinix, B. Long-Term Phosphorus and Potassium Application to Corn on Coastal Plain Soils. *jpa* **1996**, *9*, 88. [CrossRef]
51. US Climate Data. Climate for Florence-South Carolina and Weather Averages. Available online: https://www.usclimatedata.com/climate/Florence/south-carolina/united-states/ussc0113 (accessed on 26 March 2019).
52. Carr, T.; Yang, H.; Ray, C. Temporal Variations of Water Productivity in Irrigated Corn: An Analysis of Factors Influencing Yield and Water Use across Central Nebraska. *PLOS ONE* **2016**, *11*, e0161944. [CrossRef] [PubMed]
53. Novak, J.; Sigua, G.; Watts, D.; Cantrell, K.; Shumaker, P.; Szogi, A.; Johnson, M.G.; Spokas, K. Biochars impact on water infiltration and water quality through a compacted subsoil layer. *Chemosphere* **2016**, *142*, 160–167. [CrossRef] [PubMed]
54. Kissel, D.E.; Sonan, L. *Soil Test Handbook for Georgia, Special Bulletin 62*; University of Georgia: Athens, GA, USA, 2011; pp. 1–100.
55. Bohn, H.; McNeal, B.; O'Connor, G. *Soil Chemistry*; John Wiley & Sons: New York, NY, USA, 1979; pp. 1–329.
56. Marschner, H. *Mineral Nutrition of Higher Plants*, 2nd ed.; Academic Press: San Diego, CA, USA, 1995; pp. 1–889.
57. Karlen, D.L.; Sadler, E.J.; Camp, C.R. Dry Matter, Nitrogen, Phosphorus, and Potassium Accumulation Rates by Corn on Norfolk Loamy Sand1. *Agron. J.* **1987**, *79*, 649. [CrossRef]
58. Clemson University Soil Testing Laboratory. Soil Test Rating System. 2019. Available online: https://www.clemson.edu/public/regulatory/ag-srvc-lab/soil-testing/pdf/rating-system.pdf (accessed on 26 March 2019).

environments MDPI

Article

Differences in Microbial Communities and Pathogen Survival Between a Covered and Uncovered Anaerobic Lagoon

Thomas F. Ducey [1,*], Diana M. C. Rashash [2] and Ariel A. Szogi [1]

[1] Coastal Plains Soil, Water, and Plant Research Center, Agricultural Research Service, USDA, Florence, SC 29501, USA
[2] North Carolina Cooperative Extension Service, Jacksonville, NC 28540, USA
* Correspondence: thomas.ducey@ars.usda.gov; Tel.: +1-843-669-5203

Received: 21 June 2019; Accepted: 1 August 2019; Published: 6 August 2019

Abstract: Anaerobic lagoons are a critical component of confined swine feeding operations. These structures can be modified, using a synthetic cover, to enhance their ability to capture the emission of ammonia and other malodorous compounds. Very little has been done to assess the potential of these covers to alter lagoon biological properties. Alterations in the physicochemical makeup can impact the biological properties, most notably, the pathogenic populations. To this aim, we performed a seasonal study of two commercial swine operations, one with a conventional open lagoon, the other which employed a permeable, synthetic cover. Results indicated that lagoon fecal coliforms, and *Escherichia coli* were significantly influenced by sampling location (lagoon vs house) and lagoon type (open vs. covered), while *Enterococcus* sp. were influenced by sampling location only. Comparisons against environmental variables revealed that fecal coliforms (r^2 = 0.40), *E. coli* (r^2 = 0.58), and *Enterococcus* sp. (r^2 = 0.25) significantly responded to changes in pH. Deep 16S sequencing of lagoon and house bacterial and archaeal communities demonstrated grouping by both sampling location and lagoon type, with several environmental variables correlating to microbial community differences. Overall, these results demonstrate that permeable synthetic covers play a role in changing the lagoon microclimate, impacting lagoon physicochemical and biological properties.

Keywords: anaerobic lagoons; permeable cover; microbial communities; pathogens; *Enterococcus*; *Escherichia coli*

---

1. Introduction

Anaerobic lagoons remain the preferred option of manure treatment for confined swine production systems in the Southeastern United States. These earthen structures, utilized for both passive treatment and storage, are aimed at reducing the organic load of fresh manure and consequently, concentrating the nutrients contained within these waste materials. These nutrients, combined with anaerobic conditions, provide a suitable growth environment for a variety of microorganisms, including a number of pathogenic bacteria [1]. In the Southeastern U.S., liquid manure is collected under the barns using slotted floors and a shallow pit filled daily or weekly with the supernatant lagoon effluent. Any excess lagoon liquid not used for filling the shallow pit is land applied on spray fields during the crop season. Therefore, pathogens can be reintroduced into the barns with recycled lagoon liquid or deposited into the surrounding environment during land application of the lagoon wastewater, where they may eventually infect livestock or truck crops, thereby potentially entering the food chain [2].

While the construction of anaerobic lagoons tend to follow general engineering design criteria [2], swine operators have the discretion to add additional safeguards and management measures as long as such modifications continue to meet federal, state, and local regulations [3]. For instance, swine operations adjacent to communities may opt to employ synthetic lagoon covers for the control of ammonia and other malodorous compounds [4,5]. These covers can be permeable (e.g., geotextile, foam, straw) or impermeable (e.g., plastic, wood, concrete), and despite differences in cover composition, they all serve a similar purpose—to reduce emissions. One benefit to permeable covers is their ability to allow oxygen penetration, resulting in microclimate formation at the cover/lagoon interface [6,7], and such microclimates have been documented to result in the formation of biofilms [7], enhance protozoa fauna populations [6], and alter nutrient cycling patterns [4].

Given the reliance on anaerobic lagoons by the swine industry as a waste treatment measure, significant research has been conducted into understanding pathogen fate [8], nutrient cycling [9], and emissions [10] in these systems. Many of these studies have focused on open (i.e., uncovered) lagoons, primarily because they dominate the treatment landscape. Despite research demonstrating that lagoon covers utilized in swine production reduce ammonia and malodor emissions, there remains a paucity of information regarding the microbial community composition of covered lagoons, and the potential for synthetic covers to impact pathogenic populations.

Given the lack of information on the microbial communities that populate covered lagoons, and whether these lagoons can control pathogenic populations, this study was conducted with two major objectives: (i) determine potential differences in pathogen kill rates and (ii) assess microbial community differences between a covered and uncovered lagoon. A third objective, if differences are identified in the first two objectives, is to determine the relationship between environmental factors and the noted differences between the two types of lagoon systems.

2. Materials and Methods

2.1. Site Description and Sample Collection

Two commercial swine finishing operations were chosen for this study. The first operation, supporting between 2100 and 2200 animals per cycle, had an uncovered 0.55 ha lagoon, while the second operation, supporting between 1200 and 1500 animals per cycle, functioned with a synthetic, permeable membrane covering the 0.4 ha lagoon, details of which have been previously described [6]. The covered lagoon operation employed a flush tank recirculation system, while the open lagoon operation employed a shallow pit with a pull-plug flushing system for moving waste out of the house. Samples were collected seasonally, starting with a spring sampling in April of 2017, and ending with a winter sampling in February of 2018. Samples from both lagoons and houses were performed in triplicate. For the uncovered house, samples were collected from the recirculation pump, while samples collected from the covered house were collected inside, during the flushing event. Lagoon samples were collected from the top of the water column at three separate locations.

2.2. Sample Analysis

Dissolved oxygen and temperature were recorded on site using a YSI ProODO optical dissolved oxygen meter (YSI Incorporated, Yellow Springs, OH, USA) prior to transport and storage of lagoon liquid samples on ice. Additional wastewater analyses, which included total suspended solids (TSS), volatile suspended solids (VSS), pH, ammonium (NH_4-N), and total Kjeldahl-N (TKN) were performed according to Standard Methods for the Examination of Water and Wastewater [11]. Anions (Cl, SO_4-S, NO_3-N, NO_2-N) were measured by chemically suppressed ion chromatography using a Dionex 2000 Ion Chromatograph according to ASTM Standard Method D4327-11 [12], while cations (Ca, K, Mg, and Na) were measured according to ASTM Standard Method D6919-09 [13].

2.3. Pathogen Detection

Escherichia coli, fecal coliforms, and *Enterococcus* sp. were enumerated on CHROMAgar E. coli (CHROMagar, Paris, France), mFC (Sigma, St. Louis, MO, USA), and mE (Becton Dickinson, Franklin Lakes, NJ) agar, respectively. To determine colony-forming units (CFU), wastewater samples were serially diluted in sterile phosphate-buffered saline (PBS) and spiral plated in triplicate on the corresponding plates. All incubations were done aerobically, at temperatures and times as follows: *E. coli* at 37 °C for 24 h; fecal coliforms at 44.5 °C for 24 h; and *Enterococcus* sp. at 37 °C for 48 h. Due to the variability in suspended solids from sample to sample, all CFU were adjusted per gram of volatile suspended solids (CFU/gVSS) prior to log10 normalization for statistical analysis purposes.

2.4. DNA Extraction

A total of 2 mL of each wastewater sample was set aside for DNA extraction using a Qiagen Allprep PowerViral DNA/RNA Kit (Qiagen Sciences Inc, Germantown, MD). A total of 200 µL of each sample was used per extraction using protocol modifications designed to extract DNA only (no RNA) from wastewater and manure samples (i.e., no β-mercaptoethanol added to buffer solutions, and DNase steps skipped). The remaining wastewater samples were archived at −80 °C. DNA purity was determined by absorbance at 260 and 280 nm using a spectrophotometer, and quantity was determined fluorometrically using a Qubit dsDNA assay kit (ThermoFisher Scientific, Waltham, MA, USA).

2.5. Deep 16S sequencing and Analysis

Deep 16S sequencing of the V3–V4 region was performed on an Ion Torrent PGM sequencer, using a 316v2 chip and Hi-Q View sequencing reagents. Barcoded bacterial 341F (5′-CCTAYGGGRBGCASCAG-3′) and 806R (5′-GGACTACNVGGGTWTCTAAT-3′), and archaeal ARC787F (5′-ATTAGATACCCSBGTAGTCC-3′) and ARC1059R (5′-GCCATGCACCWCCTCT-3′) primers were designed according to the Ion Amplicon Library Preparation Fusion Methodology (Life Technologies, Carlsbad, CA, USA), and included 12 base pair error-correcting Golay barcodes [14]. Primers were synthesized by Integrated DNA Technologies (IDT, Coralville, IA, USA). Individual amplicon libraries for bacterial and archaeal community analysis were generated by PCR using the following protocol: activation of enzyme at 94 °C for 3 min, followed by 40 cycles of denaturation at 94 °C for 30 s, annealing at 58 °C for 30 s, and elongation at 68 °C for 45 s. Amplicons were quantified using a Qubit Fluorometer (Invitrogen, Carlsbad, CA, USA), quality controlled on an Agilent 2100 BioAnalyzer (Agilent, Santa Clara, CA, USA), and amplicons from each sample were mixed in equimolar amounts prior to sequencing.

Full-length forward- and reverse-direction sequencing libraries for each sample were verified for read quality, assembled, and analyzed using the Ion Reporter v5.10 platform and metagenomics workflow (ThermoFisher Scientific, Waltham, MA, USA). Operational taxonomic units (OTUs) were assigned at a cutoff of 97% for genus identification using the curated MicroSEQ 16S v2013.1 and Greengenes v13.5 [15] reference libraries. For determination of substrate utilization for methanogenesis, archaeal families were sorted into three groups: acetoclastic, hydrogenotrophic, and methylotrophic. *Methanosaetaceae*, *Methanosarcinaceae*, and *Methermicoccaceae* were classed as acetoclastic. The *Methanomassiliicoccaceae* was classed as methylotrophic. The remaining were classed as hydrogenotrophic.

2.6. Statistical Analysis

All statistical analyses were performed using Minitab 17 (Minitab Incorporated, State College, PA). Analysis variance (ANOVA) was conducted using the general linear model, with pairwise comparisons using Fisher's Least Square Difference Method (LSD); difference between any two means was considered significant with $p < 0.05$. Regressions of bacterial CFUs (log10 CFU/gVSS) with environmental variables were performed using a linear model. Non-metric multidimensional scaling (NMS) of microbial community population data was performed in PC-Ord v.6 (MJM Software Design, Gleneden Beach, OR, USA).

3. Results and Discussion

3.1. Wastewater Characteristics

Wastewater physicochemical characteristics are summarized in Table 1. Results are consistent with the wastewater properties of other swine-studied anaerobic lagoons [8,9]. Seasonal effects were documented for temperature, with summer samples showing significantly higher ($p < 0.05$) temperatures than other samples, and dissolved oxygen, with spring samples (0.74 ± 0.07 SE mg L^{-1}) demonstrating significantly higher dissolved oxygen (DO) levels ($p < 0.05$) than fall samples (0.37 ± 0.01 SE mg L^{-1}). Sampling location (i.e., lagoon vs house) was significant ($p < 0.05$) for pH, temperature, TSS/VSS, and TKN. Lastly, when examining the lagoons, the type of system (i.e., open versus covered) demonstrated significant differences ($p < 0.05$) for pH and TKN. Covered lagoons (1009 ± 24 SE mg L^{-1}) have more than double the TKN of open lagoons (473 ± 44 SE mg L^{-1}), this may be explained by the TKN levels originating in the animal houses that feed into those lagoons. TKN in the house feeding into the covered lagoon had mean TKN levels of 3632 ± 339 SE mg L^{-1}, while the house feeding into the open lagoon had mean TKN levels of 1306 ± 211 SE mg L^{-1}. These results differ from those of VanderZaag et al. [4], which showed no significant difference between TKN levels of a covered lagoon system when compared to an open control lagoon system filled from the same wastewater source.

Table 1. Fisher pairwise comparisons of lagoon and house physicochemical characteristics.

Season	System	Site	pH	DO	Temp	Total Suspended Soils	Volatile Suspended Solids	Total Kjeldahl Nitrogen
				mg L^{-1}	°C		mg L^{-1}	
Spring	Open	Lagoon	8.11 ^{a} [1]	0.60 ^{bcde}	20.5 ^{i}	189 ^{e}	166 ^{d}	630 ^{gh}
		House	7.24 ^{d}	0.96 ^{a}	21.5 ^{h}	3200 ^{e}	3125 ^{d}	1741 ^{e}
	Cover	Lagoon	6.92 ^{h}	0.66 ^{bcd}	20.6 ^{i}	293 ^{e}	289 ^{d}	906 ^{fg}
		House	7.33 ^{def}	0.86 ^{ab}	26.1 ^{c}	6275 ^{de}	5975 ^{cd}	2445 ^{d}
Summer	Open	Lagoon	7.86 ^{b}	0.92 ^{a}	27.3 ^{b}	149 ^{e}	124 ^{d}	276 ^{i}
		House	7.24 ^{efg}	0.25 ^{gh}	28.3 ^{a}	15,600 ^{d}	13,900 ^{c}	1850 ^{e}
	Cover	Lagoon	7.16 ^{fg}	0.53 ^{def}	27.0 ^{b}	410 ^{e}	285 ^{d}	985 ^{f}
		House	6.67 ^{i}	0.23 ^{h}	27.5 ^{b}	100,100 ^{a}	89,600 ^{a}	4823 ^{a}
Fall	Open	Lagoon	7.72 ^{bc}	0.40 ^{efgh}	20.9 ^{i}	276 ^{e}	237 ^{d}	399 ^{hi}
		House	7.35 ^{def}	0.33 ^{fgh}	22.2 ^{k}	325 ^{e}	265 ^{d}	546 ^{hi}
	Cover	Lagoon	7.28 ^{efg}	0.35 ^{fgh}	21.9 ^{gh}	483 ^{e}	410 ^{d}	1110 ^{f}
		House	7.04 ^{gh}	0.40 ^{defgh}	24.1 ^{e}	51,400 ^{b}	44,850 ^{b}	3848 ^{b}
Winter	Open	Lagoon	7.76 ^{bc}	0.55 ^{cdef}	23.3 ^{f}	467 ^{e}	413 ^{d}	588 ^{ghi}
		House	7.36 ^{def}	0.81 ^{abc}	25.6 ^{d}	1927 ^{e}	1755 ^{d}	1088 ^{f}
	Cover	Lagoon	7.41 ^{de}	0.53 ^{def}	19.1 ^{j}	427 ^{e}	373 ^{d}	1033 ^{f}
		House	7.13 ^{fgh}	0.53 ^{cdefg}	25.5 ^{d}	41,375 ^{c}	37,250 ^{b}	3410 ^{c}

[1] Means followed by the same letter are not significantly different at $p = 0.05$.

Analysis of cation and anion concentrations of swine wastewater are found in Table 2. While NO_2-N and NO_3-N were assayed, they were below detectable limits throughout the course of the study. No significant seasonal effects were noticed amongst samples, although sampling location was significant for all cations and anions detected, with significantly increased concentrations ($p = 0.05$) in swine houses. When examining the lagoon system used, SO_4-S was significantly higher ($p < 0.05$) in the open lagoons (30.8 ± 4.9 SE mg L^{-1}) as compared to covered lagoons (4.5 ± 0.6 SE mg L^{-1}); conversely, NH_4-N was significantly increased ($p < 0.05$) in covered lagoons (858 ± 11 SE mg L^{-1}) as compared to open lagoons (379 ± 38 SE mg L^{-1}). As already noted for TKN, the higher NH_4-N concentrations in the covered lagoon most likely were a consequence of higher N loading in the more concentrated wastewater derived from the house feeding into it.

Table 2. Fisher pairwise comparisons of lagoon and house anions and cations (mg L^{-1})[1].

Season	System	Site	Cl	NH_4-N	PO_4-P	SO_4-S	Ca	K	Mg	Na
Spring	Open	Lagoon	491.8 ^{f2}	490.0 ^{h}	18.7 ^{efg}	41.3 ^{c}	74.5 ^{g†}	720.1 ^{fgh}	29.6 ^{f}	254.4 ^{e}
		House	750.8 ^{c}	1182.2 ^{c}	33.3 ^{cdef}	8.4 ^{d}	123.6 ^{bcd}	983.8 ^{c}	34.9 ^{cdef}	389.8 ^{b}
	Cover	Lagoon	444.6 ^{g}	842.4 ^{de}	94.0 ^{b}	5.4 ^{d}	113.7 ^{cd}	658.5 ^{ghi}	57.2 ^{cd}	214.9 ^{f}
		House	918.3 ^{a}	1420.0 ^{b}	47.5 ^{cd}	63.8 ^{b}	82.3 ^{efg}	1363.9 ^{b}	43.7 ^{cdef}	431.2 ^{a}
Summer	Open	Lagoon	407.5 ^{h}	204.1 ^{j}	7.9 ^{g}	40.0 ^{c}	48.4 ^{h}	598.4 ^{i}	35.9 ^{def}	195.4 ^{f}
		House	543.0 ^{e}	639.8 ^{g}	10.5 ^{fg}	9.3 ^{d}	74.4 ^{g}	831.1 ^{de}	51.6 ^{cdef}	259.9 ^{e}
	Cover	Lagoon	440.7 ^{gh}	807.7 ^{ef}	8.4 ^{g}	6.5 ^{d}	26.1 ^{i}	644.1 ^{hi}	33.8 ^{ef}	206.4 ^{f}
		House	665.8 ^{d}	754.2 ^{f}	25.5 ^{cdefg}	2.9 ^{d}	191.3 ^{a}	1038.9 ^{c}	60.2 ^{cd}	345.1 ^{c}
Fall	Open	Lagoon	423.6 ^{gh}	312.5 ^{i}	16.5 ^{fg}	39.0 ^{c}	65.2 ^{gh}	666.3 ^{ghi}	46.9 ^{cdef}	203.6 ^{f}
		House	489.7 ^{f}	426.3 ^{h}	21.5 ^{defg}	32.7 ^{c}	85.9 ^{efg}	758.6 ^{ef}	52.9 ^{cdef}	239.0 ^{e}
	Cover	Lagoon	430.0 ^{gh}	885.1 ^{d}	39.5 ^{cde}	4.0 ^{d}	103.0 ^{de}	708.8 ^{fgh}	55.9 ^{cde}	198.0 ^{f}
		House	828.3 ^{b}	1580.4 ^{a}	100.8 ^{b}	118.6 ^{a}	198.6 ^{a}	1562.5 ^{a}	104.2 ^{b}	424.9 ^{a}
Winter	Open	Lagoon	424.8 ^{gh}	509.2 ^{h}	17.8 ^{efg}	2.8 ^{d}	77.8 ^{fg}	690.4 ^{fgh}	44.6 ^{cdef}	200.3 ^{f}
		House	579.7 ^{e}	886.2 ^{de}	46.3 ^{cd}	BDL[3]	101.4 ^{def}	883.7 ^{d}	62.2 ^{c}	293.4 ^{d}
	Cover	Lagoon	433.3 ^{gh}	898.0 ^{de}	46.7 ^{cd}	2.2 ^{d}	134.8 ^{bc}	736.9 ^{fg}	54.8 ^{cde}	203.9 ^{f}
		House	777.2 ^{c}	1512.3 ^{ab}	217.9 ^{a}	46.5 ^{c}	140.9 ^{b}	1627.7 ^{a}	290.5 ^{a}	422.7 ^{a}

[1] F, NO_2-N, and NO_3-N were below detectable limits (<2 mg/L); [2] Means followed by the same letter are not significantly different at $p = 0.05$; [3] BDL, below detectable limits (<2 mg/L).

3.2. Pathogen Reduction

Fecal coliforms, *E. coli*, and *Enterococcus* sp. were identified and enumerated in all samples (Supplementary Table S1). The highest rates of enumeration were found in animal houses, and at no point were CFU rates higher in a lagoon when compared to its respective house.

3.2.1. House vs. Lagoon

Comparisons between animal houses and their respective lagoon can be found in Figure 1. Differences in bacterial levels (Figure 1; Supplementary Table S1) between the houses and wastewater lagoons demonstrate that transfer of wastewater from the houses to the lagoons results in significant reductions to all three bacterial indicators measured. Given that all CFUs were adjusted based on volatile suspended solid levels, these reductions are independent of the solids concentration of the wastewater. Significant relationships ($p < 0.05$) were observed between fecal coliforms, *E. coli*, and *Enterococcus* sp. with pH, N (TKN; NH_4), chloride, K, and Na. These chemical properties demonstrated significantly higher concentrations in the houses as compared to the lagoons (Tables 1 and 2). These results are supported by Viancelli et al. [16] that similarly documented reductions in total coliforms and *E. coli* after movement of swine manure to anaerobic lagoons, a result that may be due to reductions in organic material leading to increased competition for resources.

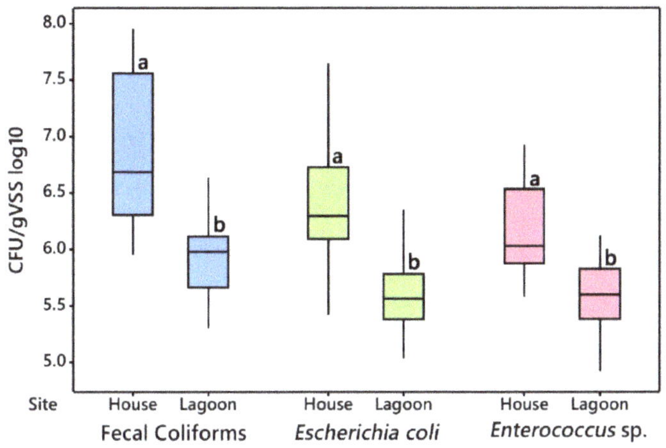

Figure 1. Comparison of colony-forming units (CFU)/gVSS $\log_{10}$ values between animal houses and lagoons, for fecal coliforms (blue), Escherichia coli (green), and Enterococcus sp. (red). Means followed by the same letter are not significantly different at $p = 0.05$.

3.2.2. Open vs. Covered Lagoon

Comparisons between open and covered lagoons can be found in Figure 2. Fecal coliform densities in the open lagoon ranged from 5.41 to 6.35 CFU/gVSS $\log_{10}$ to 5.73 to 7.04 CFU/gVSS $\log_{10}$ in the covered lagoon. *E. coli* ranged from 5.13 to 5.83 CFU/gVSS $\log_{10}$ in the open lagoon, to 5.34 to 6.47 CFU/gVSS $\log_{10}$ in the covered lagoon. *Enterococcus* sp. counts ranged from 4.88 to 5.94 CFU/gVSS $\log_{10}$ in the open lagoon, to 5.44 to 5.86 CFU/gVSS $\log_{10}$ in the covered lagoon. The CFU counts are listed in Supplementary Table S1.

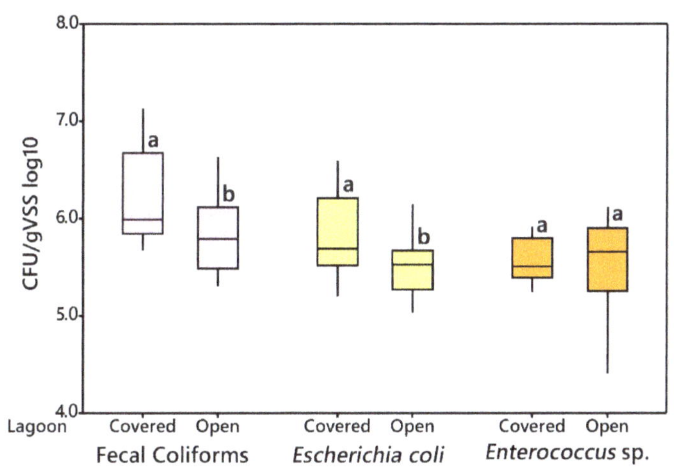

Figure 2. Comparison of CFU/gVSS $\log_{10}$ values between the covered and open lagoon, for fecal coliforms (blue), Escherichia coli (green), and Enterococcus sp. (red). Means followed by the same letter are not significantly different at $p = 0.05$.

Analysis of variance examining seasonal, site specific, and sampling location effects demonstrated that all three variables play significant roles in pathogen reduction. While seasonal patterns emerged in CFU counts for all three measured bacterial populations, with highest densities tending to be in the summer samplings, and the lowest densities observed during the winter, only fecal coliform counts demonstrated a significant relationship with temperature (r^2 = 0.485, p = 0.003). Examination between pathogen counts and physicochemical characteristics revealed significant relationships between pH and fecal coliforms (r^2 = 0.404, p = 0.008), *E. coli* (r^2 = 0.577, p = 0.001), and *Enterococcus* sp. (r^2 = 0.248, p = 0.05). For *E. coli*, significant relationships between TKN (r^2 = 0.261, p = 0.04), chloride (r^2 = 0.471, p = 0.003), potassium (r^2 = 0.272, p = 0.04), and sodium (r^2 = 0.394, p = 0.009) were also identified. No further influences on bacterial counts by physicochemical parameters were noted.

Additionally, for the lagoons, it appears that the addition of a cover had a significant impact on fecal coliform and *E. coli* levels, resulting in increased CFUs. It is possible that these higher bacterial densities in the covered lagoon may be due to solar radiation. Reductions in solar radiation have been demonstrated to result in increased bacterial counts [8], and may be a contributing factor in the increased bacterial counts in the studied covered lagoon. For the open lagoon, pH was significantly higher as compared to its covered counterpart (Table 1), and Curtis et al. identified that pH levels over 7.5, combined with sunlight, reduced fecal coliform levels [17]. *E. coli* thrive in a relatively neutral pH range, up to around pH 7.75, after which they begin to become stressed [18]. It should be noted that the open lagoon had pH ranges at or above this 7.75 pH value and could be contributing to the lower CFU counts observed. While increased pH may contribute to reductions in bacterial pathogens, it also results in increased ammonia volatilization. To counter this phenomenon, acidification is employed to reduce ammonia emissions from swine wastewater [19], and if modest reductions in pH (by two to three units) can also achieve significant pathogen reduction levels, it may provide producers with an additional means to reduce environmental impacts. This was demonstrated by Odey et al. who utilized lactic acid fermentation to inactivate fecal coliforms in human fecal sludge by reducing the pH to 3.9 [20]. Additionally, *E. coli* is considered a major reservoir of antibiotic resistance genes [21], so any employable means to reduce *E. coli* CFUs could prove to be a treatment capable of disrupting the cycle of antimicrobial resistance of animal origin.

3.3. Microbial Community Composition

Microbial community analysis using non-metric multidimensional scaling (Figures 3 and 4) revealed significant differences in the bacterial and archaeal population structures of the open and covered systems. While the samples taken from the lagoon and house of the open system showed a high degree of similarity, as evidenced by their overlapping groupings (Figures 3 and 4), the lagoon and house from the covered system neither overlapped with the open system, or each other. This pattern was similar in both the bacterial and archaeal NMS plots, and indicate larger differences in the population structure of the covered system. A number of environmental relationships correlate with these differences for bacterial populations (Figure 3), and are as follows: along the first axis, pH (r^2 = 0.354), TSS/VSS (r^2 = 0.467), TKN (r^2 = 0.513), K (r^2 = 0.532), and Na (r^2 = 0.338); and along the second axis, chloride (r^2 = 0.316), pH (r^2 = 0.266), and Na (r^2 = 0.258). Both TKN and suspended solids have been previously demonstrated to correlate with bacterial community structure [22]. Archaeal populations (Figure 4) correlated with several environmental variables along the first axis, Ca (r^2 = 0.468), K (r^2 = 0.468), and TKN (r^2 = 0.422). Calcium has been demonstrated to impact anaerobic digestion at concentrations as low as 100 mg L^{-1} [23], while potassium has been reported as toxic to acetate-utilizing methanogens [24].

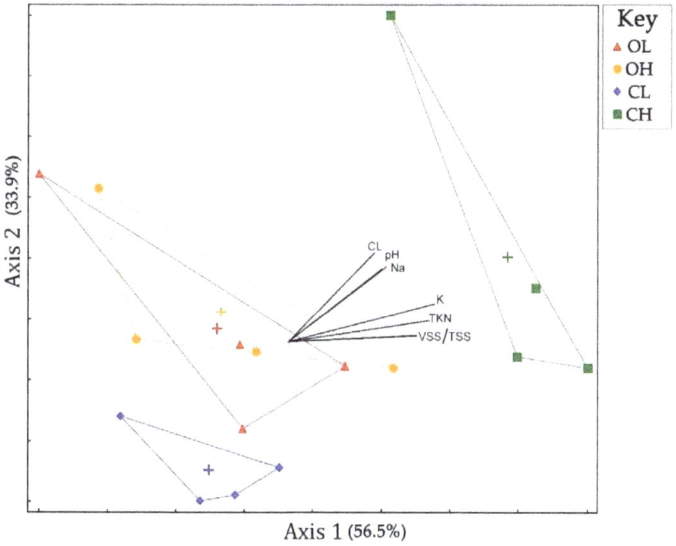

Figure 3. Non-metric multidimensional scaling (NMS) plot of microbial communities (based on relative abundances of bacterial families identified). Only explanatory environmental variables with a combined $r^2 > 0.45$ for both axes are included as vectors. Centroid for each group is marked by (+). O = open; C = cover; L = lagoon; H = house.

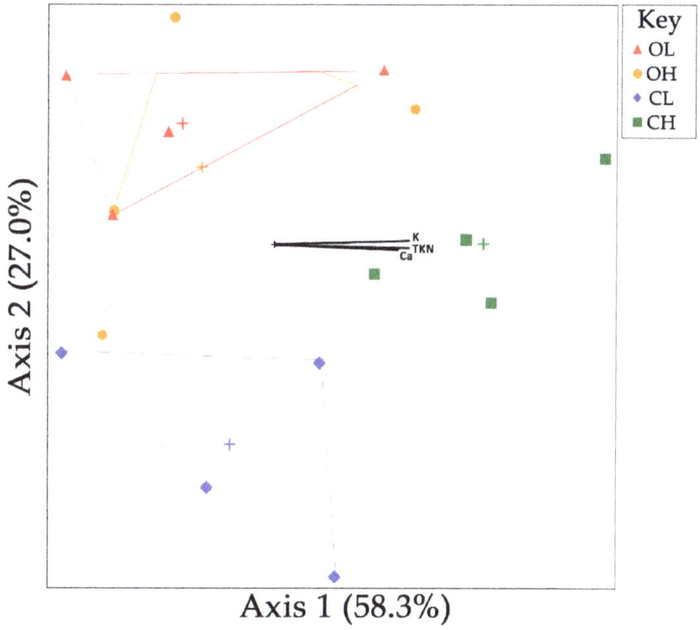

Figure 4. Non-metric multidimensional scaling (NMS) plot of archaeal communities (based on relative abundances of archaeal families identified). Only explanatory environmental variables with a combined $r^2 > 0.45$ for both axes are included as vectors. Centroid for each group is marked by (+). O = open; C = cover; L = lagoon; H = house.

The bacterial populations of both the covered and open lagoons demonstrate similarity to lagoons previously reported [22,25]. Of the 231 families identified in the 16 waste samples collected over the course of the study, using the universal bacterial primer set, only 22 bacterial families (9.5%) were represented in all 16 samplings. However, these bacterial families account for an average of 62.2% (± 4.2% SE; range 33.1% to 86.3%) of the OTU sequences classified in each sample (Figure 5; Supplementary Material Table S2). A total of 34 (14.7%) bacterial families were represented in all 8 lagoon samplings (see Supplementary Material Table S2). Of these 34 families, several were previously reported as being ubiquitous in analyzed anaerobic swine lagoons [22], such as *Ruminococcaceae*, *Chlostridiaceae*, *Lachnospiraceae*, *Peptostreptococcaceae*, and *Synergistaceae*. One noticeable difference is that while previous studies demonstrated high levels of *Chromatiaceae*, in this particularly study, this family went unidentified in the covered lagoon samples. The *Chromatiaceae*, also referred to as purple sulfur bacteria, rely primarily on phototrophic growth [26], and their growth in open lagoons is often quite evident, particularly when the lagoons adopt a purplish to red hue [27]. This family accounted for approximately half the OTU sequences for the open lagoon in the spring (56.3%) and summer (55.0%) samplings (see Figure 5). The greenish tint of the covered lagoon samples compared to the purplish tint of the open lagoon samples during sampling lent support to these findings. These findings potentially correlate with the significantly higher levels of SO_4-S in the open lagoons as compared to the covered lagoons, due to sulfate oxidation by purple sulfur bacteria [28]. These results are similarly reflected in the identification of *Desulfomicrobiaceae*, a family of sulfate reducers [29], only in samples taken from the open lagoon system. Additional sulfate reducers belonging to the families *Desulfobacteraceae* (8 of 8), *Desulfobulbaceae* (8 of 8), and *Desulfovirbrionaceae* (7 of 8) were found in a majority of all lagoon samples [30]. Additionally, while primers 341F and 806R were designed as bacterial-specific, they have been known to pick up archaeal sequences [31]. This led to the identification of the *Methanobacteriaceae*, an archaeal family of hydrogenotrophic (H_2/CO_2) methane (CH_4) producers, and the *Methanosaetaceae*, a family of archaeal acetoclastic methanogens, both of which were found in all 16 samples. The identification of these two archaeal families is of particular importance given the interest of the pork industry to use impermeable lagoon covers to trap methane for energy production in their "manure-to-energy" program initiative [32].

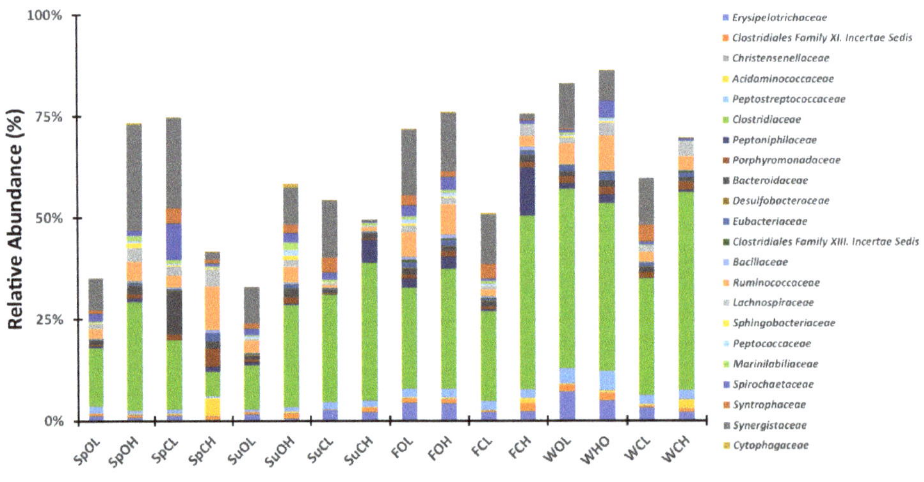

Figure 5. Bacterial community structure, shown as relative abundance. The legend listing selected bacterial families is displayed to the right of the chart. Samples are distinguished by columns. Sp = spring; Su = summer; F = fall; W = winter; O = open; C = cover; L = lagoon; H = house.

A closer look at the archaeal community composition (Figure 6; Supplementary Table S3), using archaeal-specific primers confirmed the presence of OTUs classified to *Methanobacteriaceae* and *Methanosaetaceae*, as well OTUs classified to five other methanogenic archaeal families, identified in all 16 samples: *Methanospirillaceae*, *Methanomicrobiaceae*, *Methanosarcinaceae*, *Methanocorpusculacea*, and the methylotropic *Methanomassiliicoccaceae*. The family *Thermofilaceae* was also identified in all 16 samples, bringing the number of families found in all 16 samples up to eight. When looking at just the eight lagoon samples, a total of 11 families were identified, and include the above-mentioned eight, as well as *Methanpyraceae*, *Methanocalculaceae*, and *Thermococcaceae*. The remaining classified OTUs were assigned to families not found in all samples, and typically found in low percentages (often less than 1%). Examination of archaeal families in relation to sampling source reveals a number of associations (Supplementary Figure S4). For example, both the *Methanosaetaceae* and *Methanoregulaceae* associate closely with the closed lagoon samples, while the *Methanospirillaceae*, *Methanocorpusculaceae*, and *Methanopyraceae* closely associate with the open lagoons and houses. The *Methanobacteriaceae*, on average the most identified archaeal family across all samples (Mean: 34.5%; SE: ± 6.0%), associate most closely with the closed house samples.

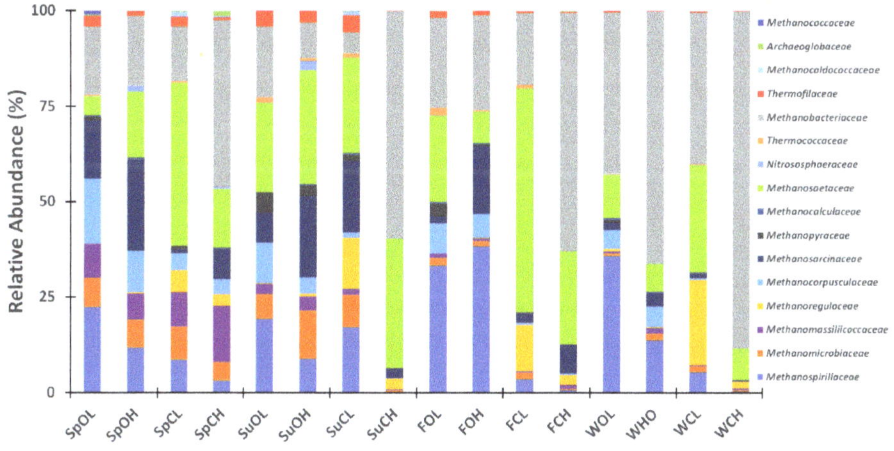

Figure 6. Archaeal community structure, shown as relative abundance. The legend listing specific families is displayed to the right of the chart. Samples are distinguished by columns. Sp = spring; Su = summer; F = fall; W = winter; O = open; C = cover; L = lagoon; H = house.

Of all the archaeal families identified, a majority of the OTUs corresponded to three, *Methanospirillaceae*, *Methanobacteriaceae*, and *Methanosaetaceae*, with the first two classified as being hydrogenotrophic methanogens, and the third classified as acetoclastic methanogens. Overall, our results demonstrate that while hydrogenotrophic methanogens make up the largest segment of methanogens in the two systems studied, acetoclastic methanogens also make up a sizeable portion of the overall methanogenic community. Seasonally, methylotrophic methanogenic OTUs were highest in the spring, acetoclastic methanogenic OTUs were highest in the summer and fall, and hydrogenotrophic methanogenic OTUs peaked in the winter (Figure 7). These OTUs point to both the open and covered lagoons as having significant potential for methane production—a process likely supported by the anaerobic conditions of the lagoons and houses, as indicated by low DO measurements (Table 1).

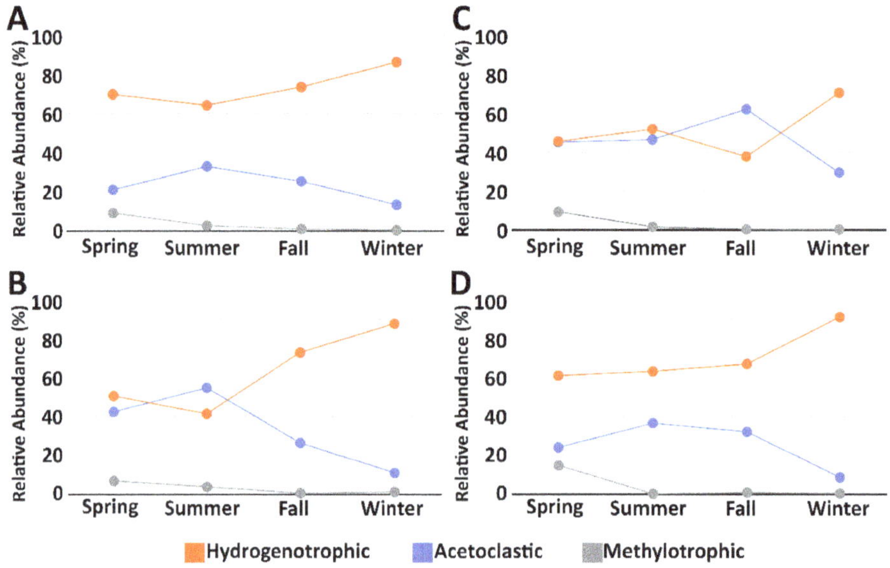

Figure 7. Relative abundance of classified operational taxonomic units (OTUs) potentially involved in methanogenic pathways. (**A**) Open Lagoon; (**B**) Open House; (**C**) Covered Lagoon; (**D**) Covered House.

4. Conclusions

While synthetic covers provide an option for swine producers to reduce odor emissions from anaerobic lagoons, there have been few studies focused on analyzing the biological responses to the microclimates generated at the cover/lagoon interface. Several wastewater physicochemical characteristics demonstrated seasonal variation, while additional differences were seen in comparisons by sampling site (lagoon vs. house) and by the type of lagoon system employed (open vs. covered). Fecal coliforms, *E. coli*, and *Enterococcus* sp., all demonstrated significant relationships with pH. When looking at fecal coliforms and *E. coli*, significant differences in CFU were identified seasonally, by sampling site, and type of lagoon system. Enterococcus sp. were unaffected by the lagoon system employed.

Microbial community analysis identified over 200 bacterial families, with 10.4% represented in all 16 samples, and an additional 19 archaeal families were identified, with eight represented by OTUs in all 16 samples. Evidence for the potential for sulfate-reduction, acetoclastic, hydrogenotrophic, and methylotrophic methanogenesis in the lagoons was demonstrated by the identification of microbial populations responsible for those processes across all lagoon samples. The in-depth sequence analysis of methanogenic communities indicates the potential for—or presence of—methane production from these anaerobic lagoons, although inhibitory concentrations of several nutrients such as Ca and K, need to be accounted for if lagoons are converted for biogas capture with impermeable covers.

Supplementary Materials: The following are available online at http://www.mdpi.com/2076-3298/6/8/91/s1, Figure S4: NMS ordination plot, as seen in Figure 4, demonstrating lagoon and house community structure in relation to individual archaeal family relative abundances, Table S1: Fisher pairwise comparisons of lagoon and house pathogen levels (CFU/gVSS log10), Table S2: Relative abundances of OTUs identified using universal bacterial primer set, presented as relative abundances (%) Only bacterial families are counted in Figure 5 and discussion involving bacterial family identification, Table S3: Relative abundances of OTUs identified using archael primer set, presented as relative abundances (%).

Author Contributions: Individual contributions were as follows: conceptualization and methodology, T.F.D.; investigation, T.F.D. and D.M.C.R.; formal analysis, T.F.D. and A.A.S.; writing—original draft preparation, TFD.; writing—review and editing, T.F.D., D.M.C.R., A.A.S.

Funding: This work was supported under USDA-ARS National Program 212.

Acknowledgments: The authors would like to thank Hannah Rushmiller and Paul Shumaker for their technical expertise. The mention of firm names or trade products does not imply that they are endorsed or recommended by the U.S. Department of Agriculture over other firms or similar products not mentioned.

Conflicts of Interest: The authors declare no conflict of interest.

References

1. Bicudo, J.R.; Goyal, S.M. Pathogens and manure management systems: A review. *Environ. Technol.* **2003**, *24*, 115–130. [CrossRef] [PubMed]
2. USEPA. *Wastewater Technology Fact Sheet: Anaerobic Lagoons*; USEPA: Washington, DC, USA, 2002.
3. USDA-NRCS. *Conservation Practice Standard: Waste Treatment Lagoon (Code 359)*; USDA-NRCS: Washington, DC, USA, 2017.
4. VanderZaag, A.; Gordon, R.; Jamieson, R.; Burton, D.; Stratton, G. Permeable synthetic covers for controlling emissions from liquid dairy manure. *Appl. Eng. Agric.* **2010**, *26*, 287–297. [CrossRef]
5. Zahn, J.A.; Tung, A.E.; Roberts, B.A.; Hatfield, J.L. Abatement of ammonia and hydrogen sulfide emissions from a swine lagoon using a polymer biocover. *J. Air Waste Manag.* **2011**, *51*, 562–573. [CrossRef]
6. Miner, J.; Humenik, F.; Rice, J.; Rashash, D.; Williams, C.; Robarge, W.; Harris, D.; Sheffield, R. Evaluation of a permeable, 5 cm thick, polyethylene foam lagoon cover. *Trans. ASAE* **2003**, *46*, 1421. [CrossRef]
7. Miller, D.N.; Baumgartner, J.W. Nitrification and denitrification potential associated with semi-permeable swine waste lagoon covers. In Proceedings of the International Symposium on Air Quality and Waste Management for Agriculture, Broomfield, Colorado, 16–19 September 2007; American Society of Agricultural and Biological Engineers: St Joseph, MI, USA, 2007; p. 14.
8. McLaughlin, M.R.; Brooks, J.P.; Adeli, A. Temporal flux and spatial dynamics of nutrients, fecal indicators, and zoonotic pathogens in anaerobic swine manure lagoon water. *Water Res.* **2012**, *46*, 4949–4960. [CrossRef] [PubMed]
9. Ducey, T.F.; Shriner, A.D.; Hunt, P.G. Nitrification and denitrification gene abundances in swine wastewater anaerobic lagoons. *J. Environ. Qual.* **2011**, *40*, 610–619. [CrossRef] [PubMed]
10. Blunden, J.; Aneja, V.P. Characterizing ammonia and hydrogen sulfide emissions from a swine waste treatment lagoon in North Carolina. *Atmos. Environ.* **2008**, *42*, 3277–3290. [CrossRef]
11. APHA. *Standard Methods for the Examination of Water and Wastewater*, 20th ed.; APHA: Washington, DC, USA, 1998.
12. ASTM. *D4327-11, Test Method for Anions in Water by Chemically Suppressed Ion Chromatography*; ASTM International: West Conshohocken, PA, USA, 2011.
13. ASTM. *D6919-09, Test Method for Determination of Dissolved Alkali and Alkaline Earth Cations and Ammonium in Water and Wastewater by Ion Chromatography*; ASTM International: West Conshohocken, PA, USA, 2009.
14. Caporaso, J.G.; Lauber, C.L.; Walters, W.A.; Berg-Lyons, D.; Lozupone, C.A.; Turnbaugh, P.J.; Fierer, N.; Knight, R. Global patterns of 16S rRNA diversity at a depth of millions of sequences per sample. *Proc. Natl. Acad. Sci. USA* **2010**, *108*, 4516–4522. [CrossRef]
15. McDonald, D.; Price, M.N.; Goodrich, J.; Nawrocki, E.P.; DeSantis, T.Z.; Probst, A.; Andersen, G.L.; Knight, R.; Hugenholtz, P. An improved greengenes taxonomy with explicit ranks for ecological and evolutionary analyses of bacteria and archaea. *ISME J.* **2011**, *6*, 610–618. [CrossRef]
16. Viancelli, A.; Kunz, A.; Steinmetz, R.L.; Kich, J.D.; Souza, C.K.; Canal, C.W.; Coldebella, A.; Esteves, P.A.; Barardi, C.R. Performance of two swine manure treatment systems on chemical composition and on the reduction of pathogens. *Chemosphere* **2013**, *90*, 1539–1544. [CrossRef]
17. Curtis, T.P.; Mara, D.D.; Silva, S.A. Influence of pH, oxygen, and humic substances on ability of sunlight to damage fecal coliforms in waste stabilization pond water. *Appl. Environ. Microbiol.* **1992**, *58*, 1335–1343. [PubMed]
18. Curtis, T. Bacterial removal in wastewater treatment plants. In *Handbook of Water and Wastewater Microbiology*; Mara, D., Horan, N., Eds.; Academic Press: Cambridge, MA, USA, 2003; pp. 477–490.
19. Fangueiro, D.; Hjorth, M.; Gioelli, F. Acidification of animal slurry—A review. *J. Environ. Manag.* **2015**, *149*, 46–56. [CrossRef] [PubMed]

20. Odey, E.A.; Li, Z.; Zhou, X.; Yan, Y. Optimization of lactic acid fermentation for pathogen inactivation in fecal sludge. *Ecotoxicol. Environ. Saf.* **2018**, *157*, 249–254. [CrossRef] [PubMed]
21. Bailey, J.K.; Pinyon, J.L.; Anantham, S.; Hall, R.M. Commensal *Escherichia coli* of healthy humans: A reservoir for antibiotic-resistance determinants. *J. Med. Microbiol.* **2010**, *59*, 1331–1339. [CrossRef] [PubMed]
22. Ducey, T.F.; Hunt, P.G. Microbial community analysis of swine wastewater anaerobic lagoons by next-generation DNA sequencing. *Anaerobe* **2013**, *21*, 50–57. [CrossRef]
23. Sharma, J.; Singh, R. Effect of nutrients supplementation on anaerobic sludge development and activity for treating distillery effluent. *Bioresour. Technol.* **2001**, *79*, 203–206. [CrossRef]
24. Schnürer, A.; Zellner, G.; Svensson, B.H. Mesophilic syntrophic acetate oxidation during methane formation in biogas reactors. *FEMS Microbiol. Ecol.* **1999**, *29*, 249–261. [CrossRef]
25. Whitehead, T.R.; Cotta, M.A. Characterisation and comparison of microbial populations in swine faeces and manure storage pits by 16S rDNA gene sequence analyses. *Anaerobe* **2001**, *7*, 181–187. [CrossRef]
26. Imhoff, J.F. *The Chromatiaceae*; Springer: New York, NY, USA, 2006; pp. 846–873.
27. Sletten, O.; Singer, R.H. Sulfur bacteria in red lagoons. *J. Water Pollut. Control Fed.* **1971**, *43*, 2118–2122.
28. Ghosh, W.; Dam, B. Biochemistry and molecular biology of lithotrophic sulfur oxidation by taxonomically and ecologically diverse bacteria and archaea. *FEMS Microbiol. Rev.* **2009**, *33*, 999–1043. [CrossRef]
29. Kuever, J.; Galushko, A. *The Family Desulfomicrobiaceae*; Springer: Berlin/Heidelberg, Germany, 2014; pp. 97–102.
30. Belila, A.; Abbas, B.; Fazaa, I.; Saidi, N.; Snoussi, M.; Hassen, A.; Muyzer, G. Sulfur bacteria in wastewater stabilization ponds periodically affected by the 'red-water' phenomenon. *Appl. Microbiol. Biotechnol.* **2012**, *97*, 379–394. [CrossRef] [PubMed]
31. Yang, S.; Phan, H.V.; Bustamante, H.; Guo, W.; Ngo, H.H.; Nghiem, L.D. Effects of shearing on biogas production and microbial community structure during anaerobic digestion with recuperative thickening. *Bioresour. Technol.* **2017**, *234*, 439–447. [CrossRef] [PubMed]
32. Smithfield Foods Announces Landmark Investment to Reduce Greenhouse Gas Emissions. Available online: https://www.smithfieldfoods.com/press-room/company-news/smithfield-foods-announces-landmark-investment-to-reduce-greenhouse-gas-emissions (accessed on 14 June 2019).

environments

MDPI

Correction

Correction: Ducey et al. Differences in Microbial Communities and Pathogen Survival Between a Covered and Uncovered Anaerobic Lagoon. *Environments*, 2019, 6, 91

Thomas F. Ducey [1,*], Diana M. C. Rashash [2] and Ariel A. Szogi [1]

[1] Coastal Plains Soil, Water, and Plant Research Center, Agricultural Research Service, USDA, Florence, SC 29501, USA; ariel.szogi@ars.usda.gov

[2] North Carolina Cooperative Extension Service, Jacksonville, NC, 28540, USA; diana_rashash@ncsu.edu

* Correspondence: thomas.ducey@ars.usda.gov; Tel.: +1-843-669-5203

Published: 24 September 2019

The authors would like to correct the published article [1]:

On page 11, Figure 7C,D should be changed from:

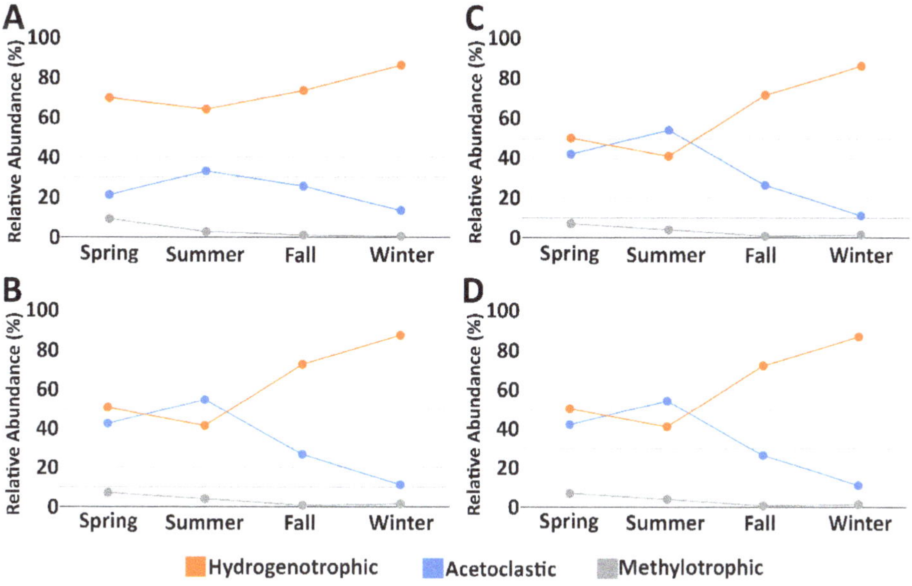

Figure 7. Relative abundance of classified operational taxonomic units (OTUs) potentially involved in methanogenic pathways. (**A**) Open Lagoon; (**B**) Open House; (**C**) Covered Lagoon; (**D**) Covered House.

to the following correct version:

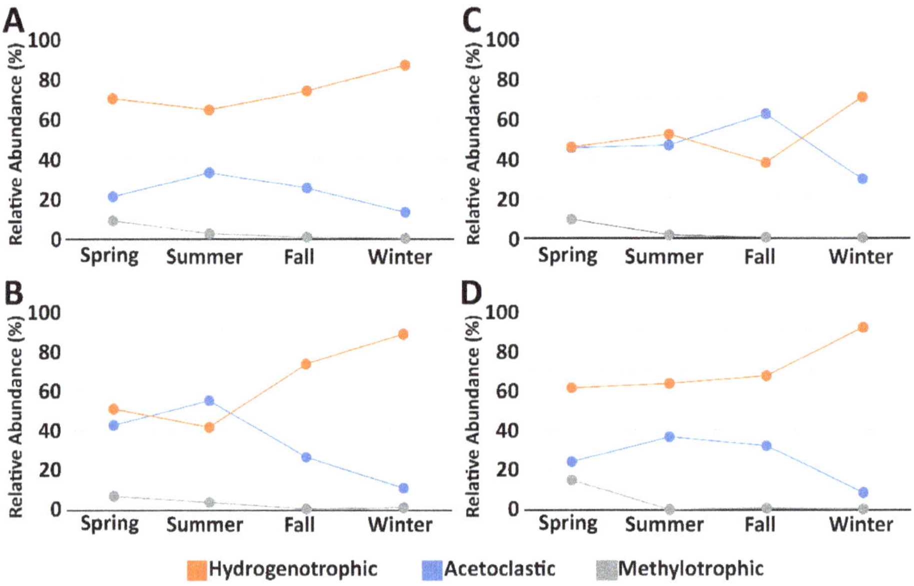

Figure 7. Relative abundance of classified operational taxonomic units (OTUs) potentially involved in methanogenic pathways. (**A**) Open Lagoon; (**B**) Open House; (**C**) Covered Lagoon; (**D**) Covered House.

The authors apologize for any inconvenience this has caused to the readers. The changes do not affect the scientific results of this paper. The manuscript will be updated, and the original version will remain online on the article webpage, with a reference to this Correction.

References

1. Ducey, T.F.; Rashash, D.M.C.; Szogi, A.A. Differences in Microbial Communities and Pathogen Survival Between a Covered and Uncovered Anaerobic Lagoon. *Environments* **2019**, *6*, 91. [CrossRef]

MDPI
St. Alban-Anlage 66
4052 Basel
Switzerland
Tel. +41 61 683 77 34
Fax +41 61 302 89 18
www.mdpi.com

Environments Editorial Office
E-mail: environments@mdpi.com
www.mdpi.com/journal/environments

www.ingramcontent.com/pod-product-compliance
Lightning Source LLC
LaVergne TN
LVHW071953080526
838202LV00064B/6738
* 9 7 8 3 0 3 9 2 1 9 5 6 8 *